VANISHING FLORA

Dionaea muscipula

INTERMER

VENUS FLYTRAP

VANISHING
FLORA

ENDANGERED PLANTS AROUND THE WORLD

TEXT

&

ILLUSTRATIONS

BY

DUGALD

STERMER

HARRY N. ABRAMS, INC., PUBLISHERS

EDITOR: *Sharon AvRutick*
DESIGNERS: *Samuel N. Antupit and Dugald Stermer*
ASSISTANT DESIGNER: *Miko McGinty*

Library of Congress Cataloging-in-Publication Data
Stermer, Dugald, 1936–
Vanishing flora: endangered plants around the world / text and
illustrations by Dugald Stermer.
p.   cm.
Includes bibliographical references (p.   ).
ISBN 0–8109–3930–4
I. Endangered plants.   I. Title.
QK86.AIS75   1995
333.95'3137—dc20   94–22873

Published in 1995 by Harry N. Abrams, Incorporated, New York
A Times Mirror Company
All rights reserved. No part of the contents of this book may be
reproduced without the written permission of the publisher

Printed and bound in Japan

Cypripedium arietinum

To Jeanie

For Crystal

CONTENTS

INTRODUCTION

God Almighty first planted a garden; and, indeed, it is the purest of human pleasures.
—Francis Bacon, "Essays, 46. Of Gardens"

Pleasures aside, the global garden would of ecological necessity have been brought into existence first among organisms, as plant life is absolutely essential to the survival of all other living things.

Today, it appears that it is our pleasure to systematically destroy that Eden. Botanists believe that due to the actions of humankind, five plant species now disappear from the wild *each day*, most without ever having been recognized, much less classified or analyzed. That totals twenty-five thousand lost by the end of the century, after which, it is estimated, the rate will increase: Up to ten species a day will disappear for the next couple of decades.

In many areas throughout the world, our attention and best efforts have been directed to the plight of endangered vertebrates, primarily mammals and birds. Belatedly, it has now become unthinkable among most reflective people to murder the remaining individuals of a rare or dwindling species, especially one we find to be attractive, symbolic, or in some way anthropomorphically romantic. When our sights are set on flora, however, we are back in the age of the great auk, passenger pigeon, and Labrador duck, all of which were heedlessly slaughtered into extinction.

While the World Conservation Monitoring Center lists nearly twenty-five hundred vertebrates in imminent danger of extinction, there are an estimated thirty to sixty thousand varieties of plant life on the brink of that abyss. More than fifty wildflowers native to North America have recently disappeared, and over three thousand—or one in ten in the United States—are believed to require some sort of protection in order to survive. Of

these, only 402 are listed in the most recent United States *Federal Register* (September 1993) as being either endangered or threatened, which affords them protection under the Endangered Species Act of 1973. (It should also be noted that the same document shows that there are recovery plans for only 167 of the 402.)

"Protection," unfortunately, is a somewhat misleading term in the context of governmental response to endangered species, especially regarding plants. For example, it is illegal to kill or trade in the 207 vertebrates listed as endangered in the United States; it is unlawful worldwide to trade in the 2,464 mammals, birds, reptiles, and fishes listed as endangered by the World Wildlife Fund and the International Union for the Conservation of Nature. As the laws pertain to the 402 officially listed plants, however, it is illegal to collect (kill) them only on publicly owned land; on private property, anything goes.

This audacious double standard is a legacy from English common law, which, in part, held that the king owned the deer while the peasants owned—less taxation and tithing—the crops and trees. Nothing in the Endangered Species Act of 1973 (reauthorized in 1982 and again in 1986) can protect the rarest cactus in North America from being relocated as a houseplant, unless it grows on federal land. Furthermore, in the event that a federal development—a dam, road, power plant, or the like—is stopped because it would destroy the habitat of a protected plant, that same land can be sold to a private developer. That developer, providing his plans don't require federal permits, can then go ahead and plow the remaining flora under forever, in order to build a parking lot, theme park, condominium, or golf course. Even if federal authorization is needed for a project, the prospective builder can get around the law simply by removing any trace of the listed plant prior to applying for the permits. Local and state restrictions vary widely, as do protective measures.

About the only factors standing in the way of such a gloomy state of affairs are the efforts of private citizens who, through various organizations and lobbying groups (such as native plant societies and garden and conservation clubs), apply pressure and raise consciousness sufficient to reroute developers away from specific critical habitats. As sporadic and individual as these projects are, and though they depend largely on the talents, energies, and resources of concerned volunteers, they have been amazingly successful. More about that presently.

Congressional hearings leading to the 1982 reauthorization of the Endangered Species Act provided eloquent testimony on prevailing attitudes about the conservation of flora, both in the United States and abroad. The Department of the Interior, the agency responsible for classifying and protecting native species, attempted to introduce a priority system for the protection of organisms, defending first species and then subspecies of mammals, birds, fishes, and on down the ladder. Ferns and seed plants (vascular plants) were relegated to the thoroughly vulnerable status of eleventh and twelfth. Fortunately, this entirely misguided ploy was averted, ignoring, as it does, even basic principles of ecology. But the underlying misbegotten thinking remains.

Amateur gardeners, ignorant or heedless of the rarity of the plants they pluck from the wild in order to ornament their backyards, are easy targets for the righteous wrath of conservationists. Even more deserving are the rapacious dealers in wildflowers and other fragile but decorative plants like cacti, dug from their native habitats for commercial exploitation. Still, as reprehensible as these poachers are, they are not the major villains in our current critical situation.

The entirely alarming rate of extinction in the twentieth century is primarily caused by our ever-increasing takeover of the world's landmasses, humans rapidly displacing our plant and animal neighbors. That we do not yet know how to live on or in the sea may account for the fact that there are relatively few saltwater species in imminent danger of disappearance, excepting, of course, the great whales and a number of dolphins. However, pollution and careless mariculture, along with unbridled fishing, may soon close that gap. *Homo sapiens* as a species is simply crowding other forms of life off the planet.

In the rainforest, habitat to uncounted and unnamed endemic plant species, hundreds of thousands of acres are being cleared to plant feed for cattle, which are in turn being ground up for fast-food franchises in the United States and abroad. This is not hyperbole, but literal fact: We are turning some of the world's last great rainforests into hamburgers.

In 1977 the creation of a stone quarry near the southern tip of Africa ended life in the wild for a small member of the iris family, the peacock moraea (*Moraea loubseri*). Not discovered until the sixties, it now exists in small numbers in "captivity," carefully nurtured by botanists from seeds rescued from the blades of bulldozers (see pages 128–29). Similarly, but on a more positive note, a new subspecies, the Tiburon mariposa lily (*Calochortus tiburonensis*), was also discovered by accident a few years ago, on the other side of the globe, in northern California. In this case, the discovery was just in time to save the plant from destruction by a planned housing development (see pages 36–37).

The same environmental qualities that attract people to particular locations—temperate to tropical climate and accessible but varied terrain—also allow an immense variety of plant life to flourish. It is no coincidence that before the great population expansion of the nineteenth and twentieth centuries, Hawaii, California, and Florida —as well as Australia—were most hospitable habitats to a great number of endemic species. It follows that these same regions should and do have the least admirable records of extinctions and near-extinctions. Current estimates suggest that there are nearly six hundred endangered plants in Hawaii, over two hundred in California, and almost five hundred in Florida. Official listings show Australia with 1,931 endangered plants and New Zealand with another 228. In Europe, developed and populated for well over two centuries, over two thousand plants are listed as nearing extinction. (Because of the varying ability, funding, and political agenda of the reporting organizations, the absolute accuracy of these numbers is questionable. It is safe to assume, however, that the figures may be raised considerably without defying reality.)

Island species are particularly vulnerable to change caused by the encroachment of human and other immigrant species. In Michael Crichton's *The Andromeda Strain*, astronauts unwittingly brought an unknown lethal virus back to Earth from deep space. Science could do nothing to counter it, and as a species *Homo sapiens* was unable to adapt quickly enough to neutralize its effects. What the Andromeda Strain was to Earth, alien creatures and plants are to island species.

Ecologically speaking, islands are complete and self-contained, with their unique, complex ecosystems and high number of organisms unlike any to be found elsewhere. Those endemic species' defensive and adaptive skills were designed in reaction to threats found in their specific environment. A change in that environment can wipe them out. Mainland species, like the barn owl, the raccoon, and the peregrine falcon, however, have found it imperative to their survival to adapt to a wide variety of environmental changes wrought by humans, and for the most part they have done so successfully; consider the name "barn owl." It is unnecessary to point out to any farmer or gardener that there are countless mainland plants that seem to thrive anywhere, in any weather, particularly where they are entirely unwanted. (It is useful, if sometimes difficult, to remember that "a weed is only a plant we have yet to find a use for.")

To illustrate the delicate nature of island organisms, as well as the interdependence of plant and animal species, the tale of the dodo is exemplary. This unique flightless bird was first classified as *Didus ineptus*, an ignorant anthropomorphic slur at best. It evolved on Mauritius, an island off Africa in the Indian Ocean, where it fed on plant life close to the ground. It had no natural enemies, so it needed neither flight nor defensive fighting skills for survival. For eons, Mauritius was a dodo Eden.

Then, in the fifteenth century, when the first Dutch seafarers landed on the island, they began killing the slow-moving bird for food; this, even though they described it as "the sick bird" because of its taste. The Dutch brought with them, in addition to their lethal weapons and curious appetite, domestic animals—pigs, goats, and dogs—which quickly overran the island, competing with the dodo for food, as well as eating the birds' eggs out of their nests, close to the ground.

The island's fragile ecosystem was suddenly thrown way off kilter, and the dodo was incapable of adapting. That, along with outright slaughter, led to the quick and, at the time, unmourned disappearance of *Didus ineptus* (later to be posthumously reclassified as *Raphus cucullatus*). This was to be the first in a distressingly long line of untimely extinctions that we have been able to attribute directly to the actions of men.

The fate of a bird may well have implications for flora as well. There is, among those who study such arcane data, fascinating speculation regarding a connection between the dodo and a species of tree endemic to Mauritius.

Mauritius's *Calvaria major*, or tambalacoque tree, is becoming critically endangered, as all remaining individuals are well into their old age. The trees have produced ample seeds for the continuation of the species, but for some two hundred years they have failed to germinate. Some biologists propose that the dodo fed on the tambalacoque's fruit. The pits, which have an unusually thick, hard surface, would have passed through the bird's gizzard before exiting in excrement. It is possible that the action of the dodo's digestive system reduced the coating around the nut and helped the seed within to sprout. As Gerald Durrell wrote, in *Golden Bats and Pink Pigeons*, this theory has "more holes than a colander"; nevertheless, there are far stranger relationships in nature.

In a tiny hilly enclave in northern California, lupine are being squeezed out of their habitat by housing developments, and with them goes the mission blue, now among the rarest butterflies in the world, which feeds solely on the plant. There are stands of lupine elsewhere —but the lovely mission blue lives in ever-dwindling numbers only in these hills of San Bruno.

Perhaps it is nit-picking to point out that we, as a species, have not made exactly the same mistake twice in caring for our global home, but that we keep inventing similar abuses so that the result is the same: We refuse to learn from our errors, so, while the damage increases exponentially, the environment is given no chance for recovery.

The massive poisoning campaign carried out by ranchers against the prairie dog on the plains of North America didn't succeed in wiping out its primary target, but it did deplete the population of the rodent to the point where its predator, the black-footed ferret, has been starved nearly out of existence. And as one result, the ranchers have deprived themselves of a crucial ally in reducing the numbers of the offending prairie dog. Exemplary anecdotes such as this abound in the dismal history of environmental abuses.

Perhaps the most consequential of all is the clear-cutting of the remaining rainforests throughout the world, especially in South America and Africa. The Atlantic coastal forest of Brazil is ninety-nine percent gone, as are the forests of many of the small islands of Polynesia and the Caribbean. In all cases, not only is the covering canopy of trees destroyed, but the plants closer to the ground which depend on those trees for shade and moisture are also being wiped out.

There are no truly accurate statistics available on the extinctions involved, as most of the flora affected have not even been discovered, much less classified. But we do know this: The process will, if allowed to continue unchecked, severely alter oxygen levels throughout our atmosphere. The only question is to what extent. It will also seriously diminish diversity; indeed it already has.

All the reasons usually put forth in favor of conserving mammals, birds, fishes, and reptiles are equally applicable to flora, and there are even a few more we can add. Conservationists, both those in the scientific community and laypeople, normally make their arguments based on one or more of five overlapping perspectives: Ethical, aesthetic, economic, scientific, and ecological (this last now embracing the principle of biodiversity).

The ethical position holds that we have no moral right to exploit our fellow organisms, either for nourishment or pleasure. It is, perhaps, conceivable that *Homo sapiens* could renounce meat entirely and become vegetarian, but it is certainly not practical. Another, more powerful ethical argument is that all existing life-forms have a right to be, simply because they are. Our "speciesism"— the assertion that our environment and all its creatures and plants have been created for our use and convenience —is not only ill founded but ultimately self-defeating. Nevertheless, such an argument is unlikely to be successful in preserving edible plants or animals in a developing nation where human survival is the priority; nor is it of much use against the force of big business in the United States and other industrialized nations.

The aesthetic argument is useful only with people who are generally so disposed. That each species is a thing of unique beauty, even a work of transcendent art, and that the wondrous variety of life-forms gives existence its richness, as the creative outpouring of a people gives them their culture, is sufficient cause for its protection— such a thesis is compelling to a relative few. The fragile delicacy of a spider orchid is not nearly enough to sway a significant portion of public opinion toward conservation.

The economic argument is the least compelling. It claims, pure and simple, that conservation is good business. This would be miraculous if true, but it falls apart for two reasons. First, it does not carry within its parameters a permanent solution to the problems wrought in the long term by long-range uses of our limited natural resources, being tied to what business people are pleased to call the bottom line. And second, this argument itself buys into the very economic system that largely created the problem in the first place. The motive power that propelled mankind into the destruction of its environmental home, that of continual growth and expansion, is unlikely to have a reverse gear.

Scientists maintain that conservation is necessary because the loss of even a single organism deprives them of inestimable amounts of material for analysis. In fact, many of the species already extinct could also have provided much information of great value. There is also no doubt whatsoever that science has only just begun its investigation into the millions of other organisms, including plants, so that cynics might argue that one or two more extinctions will make little difference; this is a variation of the "what you don't know can't hurt (or help) you" school of thought.

The most compelling argument for conservation is the case based on ecological grounds. None of the first four, even taken together, is going to have much impact against a conflict based on the requirements for food, land, or shelter. Immediate and real need is going to win out over any other consideration, but the imperative posed by ecology, along with its Siamese twin, biodiversity, is the only basis upon which compromises may be found. Ecology demonstrates conclusively that nature makes no value judgments based on how mammals, birds, fishes, insects, and plants relate to *Homo sapiens*. We are all subject to the same immutable laws, without any receiving preferential treatment.

A theoretical clue to the difference between a natural extinction and an unnatural, or untimely, one is that nature provides an almost instant replacement for the former, while the latter leaves a vacuum forever. Dinosaurs, for example, were replaced by marsupials and mammals that were better equipped to cope with changing environmental conditions. The problem for us is that it takes nature hundreds of thousands, even millions of years to make its changes, far too long for us to analyze the results in time to alter our behavior.

Because our ecological impact cannot become neutral —we are simply too pervasive—we must discipline ourselves to step more lightly on the planet. Otherwise, if we continue, through overconsumption, overdevelopment, and overpopulation, to destroy an ever-larger number of life-forms, we will surely unravel the complex systems that keep us alive. That, in one of the remaining nutshells, is what biodiversity is about. John C. Sawhill, president and chief executive officer of The Nature Conservancy, puts it beautifully: "The consequences of this unraveling should deeply disturb anyone who cares about the future of life on earth. Like a tapestry that's frayed around the edges, the natural world has managed to retain its basic integrity despite the loss of thousands of species. But as species continue to vanish, the tapestry of nature will soon collapse. . . . In other words, today we're losing species perhaps ten thousand times faster than we have over the past four hundred million years."

The lesson ecology teaches us is basic: When we impel a species into extinction, it leaves a void in its wake; however, most often the manmade causes for its disappearance also remain behind, a threat to every other living being, ourselves included.

Humankind cannot do without plants. It is conceivable, however—although one would certainly not wish to try the hypothesis—that we could, for a time, continue to exist without many, perhaps even a majority, of mammals, birds, fishes, and reptiles, at least until the biosphere began to break down in unimagined ways.

But while all animals ("any of a kingdom [Animalia] of living beings typically differing from plants in capacity for spontaneous movement and rapid motor response to stimulation"—*Webster's Seventh New Collegiate Dictionary*) are almost entirely dependent on plants for day-to-day survival, the reverse is not the case, at least not nearly to the same degree. And yet by the turn of this century, we will have sent as many as twenty-five percent of all plant species into oblivion.

It might be useful to be reminded of the myriad of uses for which we exploit vascular plants, other than the absolute of relying on them to provide the very breath of life. For the vast majority of people throughout the world, plants mean food—not only cultivated crops, but many varieties of wild flora as well. Further, even our cultivated crops—rice, wheat, barley, etc.—require periodic influxes of wild strains to keep the gene pool strong and diverse, resulting in a plant more resistant to insect predation, climatic changes, and disease. Every five years, North American wheat is crossbred with a certain wild variety to keep it healthy and ensure its fecundity. This wild strain of wheat was found by accident in a small rural locale in Mexico. It could easily have gone unnoticed, as it was on the verge of being plowed under, the fate of thousands of potentially useful species.

After oxygen and food, we require plants for their medicinal qualities. Nearly all our remedies were originally derived from vascular plants, including those now made with chemical and synthetic substitutes. Even today, research into new cures and remedies nearly always begins with the study of organic material derived from flora. For example, the Antioch Dunes evening primrose (*Oenothera deltoides* ssp. *howellii*, see pages 132–33), a federally listed endangered species found only in a tiny area in northern California, has enormous potential medicinal benefits. It is the only natural source, other than human milk, of an enzyme thought to help prevent or cure heart disease, eczema, and even hangovers.

Fewer than ten percent of all known plants have been analyzed for their nutritional or medicinal properties. As we are busy polluting the oceans with all manner of toxic wastes, we are just beginning to explore the vast potential of marine plants, which might possibly provide one answer to world hunger.

All preservation implies a trade-off; the cake we consume we cannot then conserve. In the development of protective measures for species such as the elephant, the California condor, or the polar bear, the stakes are enormous. These creatures are doomed in the wild unless we find some way to set aside large reserves of their habitat. Otherwise, the best we can hope for is that our children may be able to observe them in zoos. (It should be noted that such reserves would provide critical habitat for uncountable other species as well, plants not excepted—another lesson of biodiversity.)

Such a preservation effort requires a difficult—apparently nearly impossible—commitment, especially for the elephant and other large mammals endemic to Africa and Asia. In developing nations, burgeoning populations exert massive pressure on land; it is understandable to all but the most militant conservationist why their governments are hesitant about restricting large areas as wildlife preserves.

However, the measures required for the protection of most flora are relatively inexpensive and uncomplicated. This is not to suggest that we save plants at the expense of animals, but only to put the problems and their potential solutions into perspective.

Perhaps the greatest cost in time and money spent for the salvation of plant species is in science, that of botanical discovery, research, classification, analysis, and prognosis. In other words, while the protection of animals is, in medical terms, primarily curative, with flora much of the work is preventive.

Once the specific situation regarding each endangered or threatened plant is fully understood, protective measures are fairly simple—or at least easy for one to out-line—and inexpensive. Governments, on the advice of knowledgeable botanists, would simply set aside, through outright purchase or lease, habitats for the plant. Given that most flora need but a few acres for healthy populations, including gene diversity, this is not usually a difficult matter. Of course, those acres must be either the plant's original habitat, or one like it in terms of terrain, climate, soil composition, and any other relevant conditions. High success rates of plant recovery in England, the United States, Mauritius, and elsewhere demonstrate that the gamble is a good one.

The occasional species of plant, as well as animal, may well resist all conservation efforts, and in that natural selection is playing a part. Such may be the case with the *Agave arizonica*, a species of cactus commonly called the century plant, and the California condor. In the *Red Data Book*, published by the International Union for the Conservation of Nature, *A. arizonica* is described as being "a plant on the verge of extinction from natural causes." However, "the inaccessibility of its habitat may insure survival, despite the critically low population," at least until off-road vehicles find it. On the other hand, ornithologists may be fighting a losing battle on behalf of the California condor, a large bird from prehistoric times, which seems determined to disappear despite our best efforts. We may be keeping the last of its kind alive past its time, a not-insignificant accomplishment.

With many endangered wildflowers, populations can be greatly enlarged through the efforts of interested laypersons, whose gardens, if selected for soil and climate, can become sites for productive and beautiful "captive breeding" programs. Obviously, such projects should be undertaken only after consultation with and under the tutelage of botanists who specialize in rare wild flora. However, nearly every major university and city contains at least one endangered plant program—sponsored by arboretums, native plant societies, The Nature Conservancy, even garden clubs—that invite the participation of private citizens.

Another remarkable effort, carried out largely by volunteers on a human scale, is that of seed saving. All over North America, individuals and small companies collect and disseminate seeds of a variety of disappearing and even extinct flora. Some, but not all, of these are food plants, like old-time beans and potatoes. May their kind multiply; it can only improve the gene pool. On the scientific side, cryogenic seed banks exist in arboretums and research facilities around the world. One of the best-known examples of species recovery was carried out by the wonderful Harold Koopowitz, who literally saved the peacock moraea from the blades of bulldozers in his native South Africa. While this lovely flower exists no longer in the wild, thanks to Koopowitz seeds have been sent to botanists around the world, and he has brought the flower back to bloom at the University of California at Irvine, where he now works.

Regarding the planet's few remaining rainforests, a much greater effort of public awareness and pressure on the respective governments is necessary if we are to reverse the current trend and save countless thousands of unique and irreplaceable species. The place to start is with the individual. If you are at all interested in helping out, learning more, or getting involved, please turn to the appendix at the back of this book for a partial list of organizations and institutions.

"The beauty and genius of a work of art may be reconceived though its first material expression be destroyed; a vanished harmony may yet inspire the composer; but when the last individual of a race of living things breathes no more, another heaven and another earth must pass before such a one can be again." These words were written by William Beebe (1877–1962), late of the New York Zoological Society, and they state the case with eloquent brevity. To squander what we can never recoup and cannot afford to lose is, whatever one's beliefs, a mortal sin, with us as perpetrators and our heirs the ultimate victims.

Note:
The terminology endemic to the fields of conservation, ecology, biodiversity, and botany is often confusing and contradictory, even to those most involved. In this book, we will abide, insofar as it is applicable and possible, by the definitions set forth by the International Union for Conservation of Nature. The IUCN, located in Gland, Switzerland, but with offices around the world, is considered to be the most reliable and thorough network of information on the status of the biosphere.

All the plants illustrated here are, indeed, "vanishing," but not necessarily "endangered." The IUCN Plant Red Data Book defines "endangered" as species "in danger of extinction and whose survival is unlikely if the causal factors continue operating . . . (including) taxa whose numbers have been reduced to a critical level or whose habitats have been so drastically reduced that they are deemed in immediate danger of extinction."

A vulnerable species is one "believed likely to move into the endangered category in the near future if the causal factors continue operating. Included are taxa of which most or all the populations are decreasing because of overexploitation, extensive destruction of habitat or other environmental disturbance; taxa with populations that have been seriously depleted and whose ultimate security is not yet assured; and taxa with populations that are still abundant but are under threat from serious adverse factors throughout their range."

Rare species are "taxa with small world populations that are not presently endangered or vulnerable, but are at risk. These taxa are usually localized within restricted geographical areas or habitats or are thinly scattered over a more extensive range."

The word "extinct" pretty much speaks for itself. However, the IUCN uses this category in practice for "species not found after repeated searches of known and likely areas." In addition, there are species of plants and animals which are extinct in the wild, but still exist in captivity; two such are Przewalski's horse and the peacock moraea.

Finally, according to the Red Data Book, "the Endangered and Vulnerable categories may include, temporarily, taxa whose populations are beginning to recover as a result of remedial action, but whose recovery is insufficient to justify their transfer to another category."

It should be remembered that these classifications, when they apply to specific plants, are determined by scientists and officials working with various government bureaucracies around the world. Unfortunately, in such a context, political considerations often take a higher priority than does accuracy of scientific reporting. Also, the level of expertise, not to mention funding and other resources, varies greatly from place to place.

These vagaries are, however, insignificant, considering the great mass of unassailable information we do have on the condition of life on the planet; it is certainly sufficient to claim our attention and efforts. The work of the IUCN, along with many other committed organizations and individuals, cannot be undervalued or taken for granted. It is the foundation upon which we need to rebuild our house.

THE PLATES

Acineta humboldtii

THIS anthropomorphic orchid, rare since its discovery in Ecuador in the early nineteenth century, is greatly imperiled due to the destruction of much of the Central and South American rainforest. Its existence in the wild is also threatened by orchid fanciers who collect the plant, despite the fact that it is extremely difficult to bring to bloom under cultivation. *A. humboldtii* has been seen in Venezuela and Colombia, in addition to Ecuador.

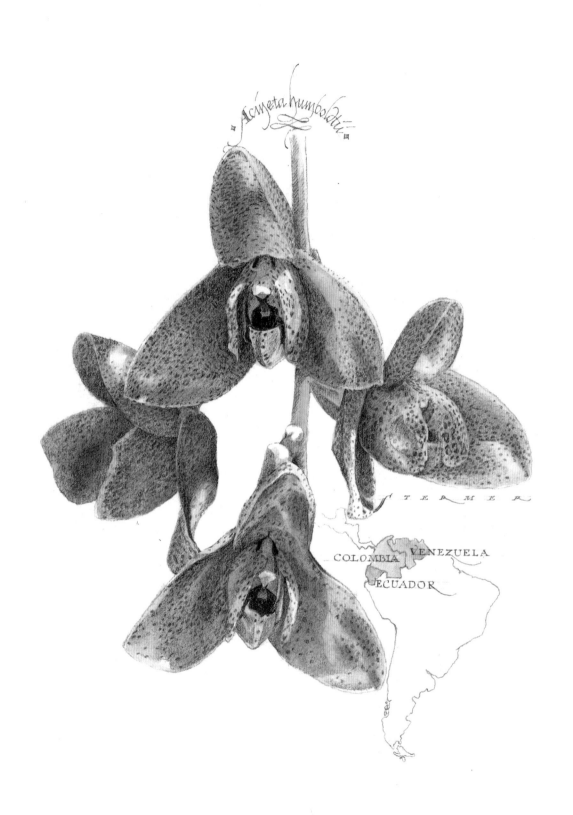

Acineta humboldtii

COLOMBIA VENEZUELA
ECUADOR

Aconitum noveboracense

NORTHERN MONKSHOOD

ACONITUM is Latin for "poisonous herb"; *noveboracense* is a latinized approximation of "New York." The plant was first named and described by botanist Asa Gray, who discovered it in 1886 in New York State. All members of the genus *Aconitum* contain poisonous alkaloids that are paralytic to the circulatory and nervous systems. They also have enormous medicinal potential; one derivative, the drug aconite, is currently being used as a sedative.

A. noveboracense is the rarest of North America's eight monkshoods. It survives in only twenty-two sites in Iowa, Wisconsin, Ohio, and New York, six of which The Nature Conservancy purchased to help protect the species. Several other sites are located in national and state parks. This plant was among the first to be granted protection under the U.S. Endangered Species Act of 1973, and it has been listed as threatened since 1978.

NORTHERN
MONKSHOOD

Aconitum noveboracense

Agalinis acuta

SANDPLAIN GERARDIA

·

FALSE FOXGLOVE

Believed to be extinct since 1940, about forty sandplain gerardia plants were suddenly discovered in September 1980, growing in a graveyard on Cape Cod. No new plants were found until four years later, when a pair of large and healthy populations was reported on Long Island. Currently, there are nine known sites: Two on Cape Cod, six on Long Island, and one in Maryland.

Although little is known about the biology of *A. acuta*, it appears that the species requires some sort of periodic habitat disruption—grass fires or other kinds of ground clearance—for its survival. Regular mowing in cemeteries preserves the open conditions sandplain gerardia requires. It is hypothesized that the suppression of wildfires, in addition to habitat destruction by development, has contributed to the species' decline.

Agalinis acuta

[false foxglove]

SANDPLAIN

GIRARDIA

STERMER

Alcimandra cathcartii

CHANGRUI MULAN

As is the case with the vast majority of endangered species—plants and animals alike—throughout the world, this native of Yunnan, China, is a victim of habitat destruction. Clear-cutting has also destroyed many stands of this tree, because it produces desirable, straight-grained wood.

Although it has been listed as endangered in the *China Plant Red Data Book* (1992) and nature preserves have been set up to protect it, the species is in dire danger. Enforcement of the laws protecting it is problematic. In years past, changrui mulan grew throughout China, Bhutan, north Burma, northeast India, and north Vietnam. Yunnan is literally its last stand.

Alcimandra cathcartii

C H A N G R U I

M U L A N

Amoreuxia wrightii

THIS lovely flower is simply too uncommon to have been given a common name. And if the shortsighted fanciers who collect it along its native range—northward from Mexico along the Rio Grande into southern Texas—have their way, it won't exist long enough to be christened.

Despite its fragile status, it has yet to be federally listed and protected.

Apopleuxia wrightii

STERMER

Astragalus lentiginosus var. *kernensis*

KERN PLATEAU LOCOWEED

·

KERN PLATEAU MILK VETCH

THIS plant, while not listed as endangered or threatened at either the state or federal level, is designated as a List Two plant for California. This means that it is rare, even at risk, within the state, but is more common elsewhere.

The Bureau of Land Management has, over the years, developed specific guidelines to be used in the protection of such rare plants as *A. lentiginosus* var. *kernensis*, including taking inventory of existing sites; determining which areas are critical to the support of rare plants; protecting those habitats by fencing, patrolling, and limiting off-road vehicle access; acquiring land designated as critical habitats; and maintaining a volunteer program to help preserve specific areas, provide technical assistance, and educate the public.

KERN PLATEAU LOCOWEED

Astragalus lentiginosus var. kernensis

Betula uber

VIRGINIA ROUND-LEAF BIRCH

First discovered in 1914 along Dickey Creek in southwest Virginia, by 1945 *B. uber* was simultaneously elevated to full species status and believed to be extinct in the wild. Then in 1975 its present, and only, population was found on the banks of Cressy Creek near Sugar Grove, in western Virginia.

Although it is federally listed as endangered, protection efforts are hampered by local vandals who, fearing government interference on private land, destroy the rare trees. Competition from neighboring vegetation has limited the birch's recovery from such wanton and shortsighted stupidity. Current restoration efforts include cultivating a healthy crop of saplings and then transplanting them to less-exposed locations.

Betula uber

STERMER

V I R G I N I A

R O U N D ~ L E A F

B I R C H

Brighamia rockii

P U A   ʻA L A

NINETY-ONE percent of
Hawaii's 1,252 native species are endemic—found no-
where else on earth—and, although the islands' Natural
Heritage Program tracks 580 rare plant taxa, only
twenty percent of the 1,252 are federally listed as
threatened with extinction. The islands' two *Brighamia*s
are among those most critically endangered.

Island species are particularly vulnerable to any disruption of or encroachment into their systems. The
Hawaiian Islands have fallen victim to invading hordes
of animals and plants to an extent unequaled on the
planet. Over two hundred years, there have been four to
five thousand alien plants introduced onto the islands,
competing for survival with the outnumbered native
species. Also wreaking havoc are the immigrant goats,
pigs, cattle, and tourists, heirs of those first brought to
the islands in 1778 by Captain James Cook.

Brighamia rockii

STERMER

PUA 'ALA

Callirhoe scabriuscula

TEXAS POPPY-MALLOW

THIS is one of the twenty plant species listed as endangered that are native to Texas. It exists on a narrow strip of land along the Colorado River in the central part of the state. The Texas Nature Conservancy and the U.S. Fish and Wildlife Service have joined forces on a recovery plan, initiated in 1987.

The Texas poppy-mallow differs from the six other species in its family in its larger leaves and wine-red flowers. It blooms for about a week, closing after sundown and then opening at daybreak. Bees and moths, which take shelter inside the flower overnight, are its most common pollinators.

callirhoe scabriuscula

TEXAS
POPPY~
MALLOW

Calochortus howellii

HOWELL'S MARIPOSA
·
HOWELL'S CALOCHORTUS

THIS mariposa has an official threatened status in its native state of Oregon, as well as a federal C2 designation, meaning that more information must be gathered before the species can be afforded protection by the U.S. government.

More than thirty plants native to Oregon have not been seen since 1960 and are considered to be extinct; eleven more face the same fate each year. The major threats to the state's flora include development, logging, and the picking of the more attractive species.

Calochortus howellii

S T E R M E R

OREGON

Only known habitat

H O W E L L'S   H O W E L L'S
C A L O C H O R T U S   M A R I P O S A

O R

Calochortus tiburonensis

TIBURON MARIPOSA LILY

In 1949, the noted botanist John Thomas Howell published his definitive *Marin Flora* (updated in 1970), listing 1,431 native species. Then in 1973 came word of one more, a newly found mariposa (shown at right in the illustration), clinging to a precarious existence on the side of soon-to-be developed Ring Mountain, literally a stone's throw from California's Highway 101, one of the most-traveled thoroughfares in the country.

News spread quickly, concerned citizens immediately formed a Save Ring Mountain committee, and they stopped the bulldozers in their tracks. Ultimately, The Nature Conservancy purchased most of the mountain, thereby protecting not only this plant but four others: *Castilleja neglecta* (Tiburon paintbrush), *Hesperolinon congestum* (Marin dwarf flax), *Calamagrostis ophitidis* (Serpentine reed grass), and *Eriogonum caninum* (Tiburon buckwheat).

Calochortus tiburonensis

1"~2"W.   4"~20"h.

Calochortus clavatus

1"~1½"W.   8"~20"h.

M A R I P O S A L I L Y

8

Calopogon pulchellus

GRASS PINK

THERE is some confusion about the specific name by which this lovely little orchid is known (such confusion is seemingly standard in botany). The generic name, about which most sources agree, is derived from two Greek words meaning "beautiful beard," referring to the fringe of yellow hair on the lip.

By any name, grass pinks are found in decreasing numbers in sphagnum bogs in a wide series of ranges, from Newfoundland south to Florida, Cuba, and the Bahamas, and west to Minnesota and even Texas. Although C. *pulchellus* is not found on federal or state lists of endangered species, the Garden Club of America lists it as being threatened with extinction in parts of Connecticut, New York, Vermont, and Virginia.

Calopogón *pulchellus*

1½" W. 4" ~ 20" h.

C. tuberósus

G R A S S P I N K S

Steiner

Camellia granthamiana

DABAOBAI SHAN CHA

THIS native of south China is threatened with immediate extinction. A recent survey indicates that only four or five clumps of this shrub remain. The *China Plant Red Data Book* (1992) recommends that the remaining sites immediately be fenced to protect the plants and states that "propagation and cultivation are urgently required." Nevertheless, it is more than a possibility that by the time this book is published, dabaobai shan cha will have completely disappeared from the wild.

Camellia granthamiana

TERMER

DABAOBAI

SHAN • CHA

Canavalia haleakalaensis

HALE-A-KA-LĀ CANAVALIA

THERE is a scientific dispute over the number of native *Canavalia* species in Hawaii: Some botanists put the number at five, others see one with four varieties, and still others define eighteen separate and distinct species.

We do know—and botanists agree—that this native of Hale-a-ka-lā on east Maui is extremely rare. Since its discovery in 1920, it has been collected fewer than six times, and in 1970 it was finally declared a new species.

In 1976 the U.S. Fish and Wildlife Service proposed listing nearly eighteen hundred plants as endangered based on a Smithsonian Institution report. Of these, 894—fifty percent—are native to Hawaii.

Canavalia haleakalaensis

J TERMER

H A L E ~ A ~ K A ~ L Ā

C A N A V A L I A

Cattleya forbesii

WHILE *Cattleya* is the best-known genus of orchid, *C. forbesii* is among the least-known species, so rare that it still has not been granted a common name.

The sixty-five species in this genus exist from Mexico south to Argentina and Peru, and they are among the most successful commercial orchids as well. There are countless hybrids resulting from crossing *Cattleya*s with orchids from related genera. Nevertheless, if the destruction of South America's rainforests is not halted, or at least abated, this particular reclusive native of Brazil may disappear before we call it by name.

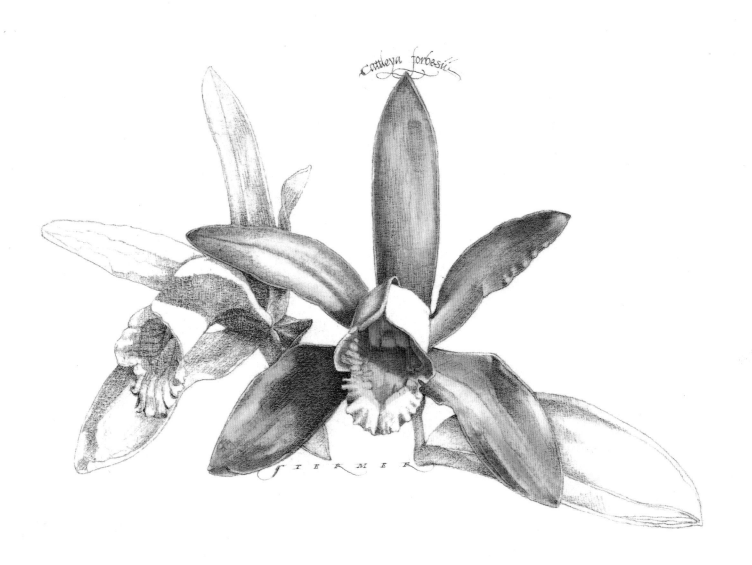

Cattleya forbesii

PIERMEK

Chrysosplenium iowense

GOLDEN SAXIFRAGE

T H E golden saxifrage, found in five sites in Iowa and Minnesota, is thought to be a relict of the Pleistocene flora that survived the last ice age. Because the region is a "driftless area," one that was not buried under ice, it supports an unusual number of rare and highly isolated populations of plants and animals. Even though *C. iowense* is not yet federally listed as endangered, state preservation programs, along with The Nature Conservancy, are protecting portions of its critical habitat.

Chrysosplenium iowense

GOLDEN · SAXIFRAGE

Cirsium loncholepis

LA GRACIOSA THISTLE

THIS is one of the rarest of
the coastal dune plants of California; nevertheless, it
remains only a potential candidate for federal and state
protection. Because the plant's neighbors include the
endangered least tern and the Belding's race of the
savannah sparrow, as well as a half a dozen plants on
the California Native Plant Society's rare and endangered list, some protection is being afforded the habitat. This includes stopping the dumping of waste-
water and crude oil that has threatened the area.

This thistle is found among the two-hundred-year-
old burial grounds and campsites of the Chumash
Indians. The area and, indeed, the thistle itself are
thought to have been named "La Graciosa" in honor of
the hospitality the Chumash showed sixteenth-century
Spanish explorer Pedro Font and his party.

Cirsium loncholepis

L A G R A C I O S A

T H I S T L E

Clematis fremontii

FREMONT'S CLEMATIS

·

FREMONT'S LEATHER FLOWER

A rare but hardy plant, *C. fremontii* seems, according to the April 1992 issue of *Missouri Conservationist*, to be holding its own against extinction. The species is not federally listed but remains on Missouri's "watch list," meaning that it has declined sufficiently to bear special attention.

Other than its primary home in east-central Missouri, *C. fremontii* is also found in three counties in north-central Kansas. *NEBRASKAland* magazine described it as "one of the most bizarre and rare flowers encountered . . . anywhere." The leaves and the petals of the flower are thick and tough, inspiring the common name.

Clematis fremontii

FREMONT'S
CLEMATIS OR
LEATHER~
FLOWER

Coryphantha minima

NELLIE CORY CACTUS

THIS tiny cactus has been listed as endangered since 1979 and remains in critical condition. Located in two separate populations near the small desert town of Marathon, Texas, it does not suffer from loss of habitat due to human encroachment. Rather, it is imperiled because of commercial cactus dealers who harvest it illegally from the wild. Thanks to the work of dedicated conservationists, the cactus is now cultivated in sufficient numbers to satisfy demand. In addition, Texas laws have been enacted to allow for the arrest and prosecution of poachers.

Coryphantha minima

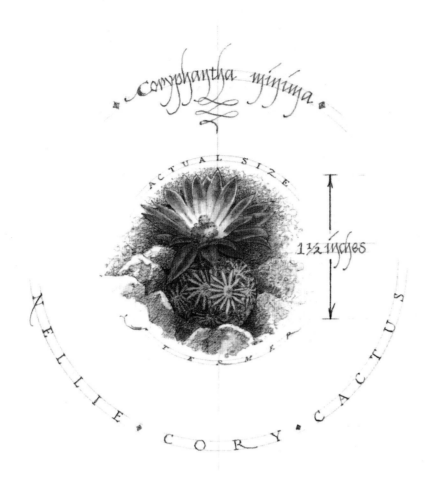

ACTUAL SIZE

1½ inches

N·ELLIE · CORY · CACTUS

Cypripedium arietinum

RAM'S HEAD ORCHID

DESPITE the fact that this orchid has an extremely wide geographical distribution —throughout North America and western China—it is still considered rare. The plant grows in thoroughly aerated soil of many kinds and succeeds in various environments, including wet, mossy swamps and dry, wooded, rocky slopes. It is even found in China at fifteen thousand feet above sea level. The flowers of this plant are among the shortest lived of all the orchids, remaining in their prime for a single day.

Cypripedium arietinum

STERMER

Cypripedium calceolus

LADY'S SLIPPER ORCHID

THIS *Cypripedium* has been picked nearly to extinction in its native European habitats and has belatedly become protected in several countries, including Austria and Switzerland. Once common throughout most of the continent, it is now extremely rare in England and thought to be extinct in Greece. People tell tales of seeing this beautiful wildflower in vases in their hotel rooms.

The exotic appearance of the flower has prompted legends, one of which concerns the goddess Venus. She was wandering through the forest when a sudden thunderstorm startled her. She ran for shelter, losing one of her golden slippers in her haste. The next day, a young shepherdess, leading her flock, saw the shoe and ran to pick it up. Just as she reached for it, however, it disappeared and was replaced by an exquisite flower in the shape of the lost slipper.

Cypripedium calceolus

LADY'S SLIPPER ORCHID

Dicliptera trifurca

ALMOST nothing is known
of this member of the Acanthus family, except that it is
a native of the higher elevation (forty-two hundred to
seventy-five hundred feet) rainforests or cloud forests
of the Cordilleras Central and Talamanca in Costa
Rica and adjacent Panama. It is to be hoped that the
rapid shrinking of its habitat will be halted, allowing
for the survival of this as well as uncounted other floral
and animal species.

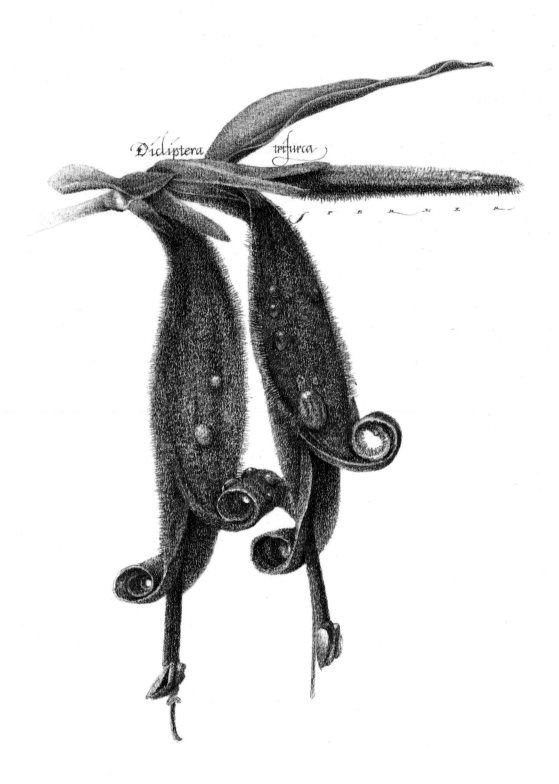

Dicliptera trifurca

Dictamnus albus

MOSES' BURNING BUSH

·

GAS PLANT

·

CANDLE PLANT

THIS is the only member
of the *Dictamnus* genus indigenous to central Europe. It
has become quite rare and is now protected in Germany
even though its range still extends from the Mediterranean across the Himalayas to northern China.

The plant was once considered a cure-all and was
widely used in herbal medicine; in some parts of its
range it is still cultivated for its therapeutic properties.
In hot weather, the aromatic oil the plant exudes is said
to burst into flame, which explains its common name.

Dictamnus albus

SEPTEMBER

Known
variously
as
GAS ~ PLANT,
CANDLE ~ PLANT, or
MOSES' BURNING BUSH

Dionaea muscipula

VENUS FLYTRAP

Among wildflowers, this plant is a true anomaly: It is severely endangered in the wild but readily available for domestic cultivation as a houseplant. In fact, the latter has largely caused the former. All of the known wild populations of *D. muscipula* are found within fifty miles of Wilmington, North Carolina, along the coastal bogs. Despite its being picked nearly to oblivion, it is still under review by the U.S. Fish and Wildlife Service for possible future protection.

The feeding habits of this carnivorous plant are a marvel of botanical engineering, accounting for its popularity in the parlor ("Watch what happens when I put a dead fly right here"). The flytrap is exactly that. The plant feeds through its leaves, which are spread out on the ground, circling the single stem. Each leaf comprises two identical blades ringed with comblike teeth and hinged along one side. The bright color of the inside of the leaf is caused by thousands of glands, which both secrete enzymes and absorb nutrients.

Dionaea muscipula

VENUS FLYTRAP

When insects or spiders are attracted by the fly-trap's sugary nectar and brush along any two of the sensitive hairs on the edge of the leaf, it slams shut, the teeth interlock, and the prey is trapped. Enzymes break down the insect's tissues, and in several days the plant digests its meal. That accomplished, the leaf again opens, the remains are blown away, and the trap is reset.

If this is a process that appeals, remember that Venus flytraps are readily available from reputable retail nurseries, so wild plants can be left in the wild.

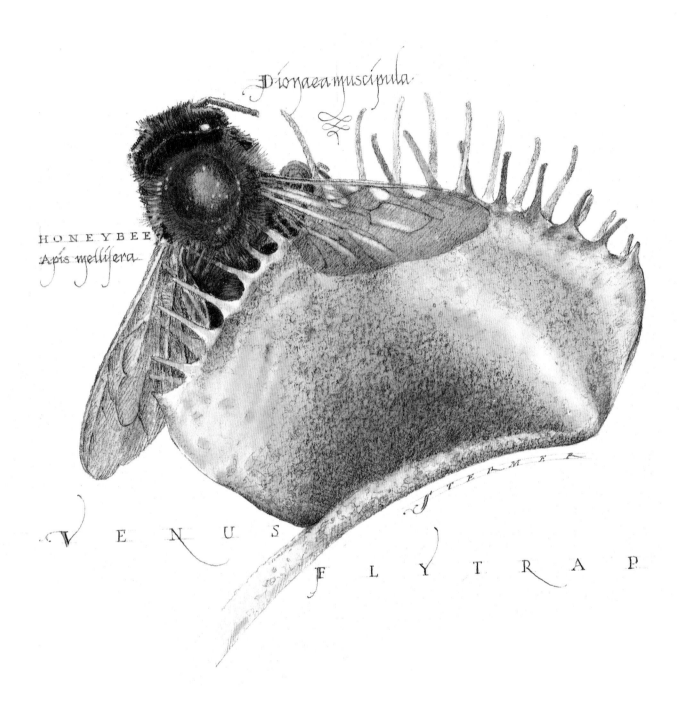

Dionaea muscipula

HONEYBEE
Apis mellifera

STELMER

VENUS FLYTRAP

Dudleya traskiae

SANTA BARBARA ISLAND
LIVEFOREVER

THIS species exists in ten colonies, with a total of approximately 230 plants, on an island just thirty-eight miles from the southern California coast. It has been listed as endangered since 1978 and receives federal protection as well as a seed propagation plan funded by the U. S. Fish and Wildlife Service. *D. traskiae* was severely depleted during the late nineteenth and early twentieth centuries by feral goats, who fed on the plants. The animals were removed in 1915, but then in 1942 New Zealand red rabbits invaded the island and continued the damage. They were finally removed twelve years later, but by then the liveforever was anything but.

Santa Barbara Island

Dudleya traskiae

L I V E F O R E V E R

Echinacea tennesseensis

TENNESSEE PURPLE

CONEFLOWER

O NE of the first plant species to be listed as endangered by the U.S. Fish and Wildlife Service, in 1979, the Tennessee purple coneflower is now found in only five sites within fourteen miles of each other in the central part of the state. While it is a close relative of *E. pallida* var. *angustifolia*, found in the midwestern United States, the two will not interbreed, and *E. tennesseensis* cannot survive harsh midwestern winters.

As part of the recovery effort, coneflowers are being cultivated at the Tennessee Valley Authority Nursery and distributed to botanical gardens throughout the state and elsewhere. Further, and most exemplary, a number of private landowners have obtained seeds and are growing the plant successfully in home gardens, helping to insure its survival.

Echinacea tennesseensis

2'~4' high

T E N N E S S E E   P U R P L E   C O N E F L O W E R

Steriner

Echinocactus horizonthalonius var. *nicholii*

NICHOL'S TURK'S HEAD CACTUS

·

EAGLE'S CLAW CACTUS

THIS is another on the short list of U.S. plants to have been listed as endangered as early as 1979. And it is threatened by the same factors that put most other desert plants at risk: Off-road vehicles, hikers, the development of limestone quarries or mines, and illegal collectors.

This rare cactus is still occasionally found in the Sonoran Desert of southern Arizona and Mexico. Fortunately, many of the plants are on land administered by the Bureau of Land Management and the Bureau of Indian Affairs—organizations that regulate use to protect some of the remaining cactus populations.

EAGLE'S CLAW CACTUS*

Echinocactus horizonthalonius var. nicholii

*NICHOL'S TURK'S HEAD CACTUS

Echinocereus engelmannii var. *purpureus*

PURPLE-SPINED HEDGEHOG CACTUS

THIS cactus was originally believed to exist in only one small location in Utah, at an elevation of twenty-nine hundred feet, and was therefore added to the federal endangered species list in 1979. Unfortunately, more recent studies indicated that it is not a distinct species after all, but instead a sporadically occurring phase of Utah's more common *E. e. chrysocentrus*. As a result, the plant may be taken off the list, leaving it defenseless against amateur and commercial collectors, not to mention off-road vehicles, heavy-footed hikers, and urban sprawl.

Echinocereus engelmannii
var. purpureus

PURPLE~SPINED

HEDGEHOG CACTUS

Erysimum capitatum var. *angustatum*

CONTRA COSTA WALLFLOWER

LISTED as endangered since 1978, the species exists in one small site, the Antioch Dunes, forty miles northeast of San Francisco, California. Its population has fluctuated between seven hundred and twenty-two hundred individual plants over the past decade.

The same fragile habitat is home to two other federally listed endangered species, the Antioch Dunes evening primrose (*Oenothera deltoides* ssp. *howellii*—see pages 132–33) and Lange's metalmark butterfly (*Apodemia mormo langei*). This area has been severely degraded and partially destroyed by industrialization, sand mining, wildfire, and human intrusion. A portion of the dunes has been purchased and added to the San Francisco Bay Area Wildlife Refuge, affording these and other rare species needed protection.

Erysimum capitatum var. *angustatum*

STERMER

CONTRA COSTA
WALLFLOWER

Eustoma grandiflorum

TULIP GENTIAN

·

PRAIRIE GENTIAN

This beautiful and rare native of northeast Colorado is a truly flamboyant wildflower, growing from one to two feet high and sporting clusters of two to six lavender, blue-purple, or pink flowers about two inches across. Although it is currently extremely uncommon in the wild, it is widely cultivated in greenhouses for both potted-plant and cut-flower commerce throughout the world. Like the Venus flytrap and many varieties of orchids, this exquisite wildflower is being bulldozed into oblivion, even as it is being commercially exploited.

Eustoma grandiflorum

TULIP PRAIRIE
or
GENTIAN GENTIAN

Franklinia alatamaha

LOST FRANKLINIA

THIS tree, with its beautiful and fragrant flowers, is the first native North American flowering plant species to have become extinct in the wild. It was discovered by pioneer botanists John Bartram and his son William in 1765, growing along the Altamaha River in Georgia. John named it after Benjamin Franklin and its native river, then spelled with an extra "a."

Eleven years later, William, taking a little time off from fighting in the Revolutionary War, went back to Georgia in search of the tree but was never able to locate it again. By around 1800, it had completely and mysteriously disappeared from its wilderness habitat.

That it survives as a cultivated tree is thanks largely to the Bartrams, who took seeds and small plants back to their own gardens, later to be given away to interested gardeners and collectors. Today three lost franklinia grow in front of the National Museum of Natural History in Washington, D.C.

« THE LOST FRANKLINIA »

Franklinia alatamaha

Gaertnera longifolia

OF the twelve species of *Gaertnera* endemic to Mauritius, *G. longifolia* is the most beautiful and easily the rarest. It exists in only about six specimens in the Perrier Nature Reserve on its native island off the east coast of Africa.

Historically, this plant flourished in the dense wooded areas of Mauritius, but as the land was cleared and put to other uses, the species disappeared from the wild. The Forestry Service of Mauritius has managed to germinate twenty-two seeds, out of which only four seedlings survive.

Gaertnera longifolia

Steiner

Gentiana crinita

FRINGED GENTIAN

BOTH the common and generic names of this striking flower refer to King Gentius of Illyria, who, according to the Roman naturalist Pliny, had discovered various medicinal uses for its roots—an emetic, a cathartic, and a tonic. Current wisdom holds that Gentius was in error and that the sole therapeutic value *G. crinita* possesses is aesthetic. It should also be noted that its generic name is in some dispute; many authorities use the one above, but others prefer *Gentianopsis crinita*.

This plant is still found, though in decreasing numbers, throughout its range, an area extending from central Maine west to Manitoba and south to Georgia, Ohio, Indiana, and northern Iowa. It is threatened primarily due to habitat destruction, but it is also being picked to death because of its unusual appearance.

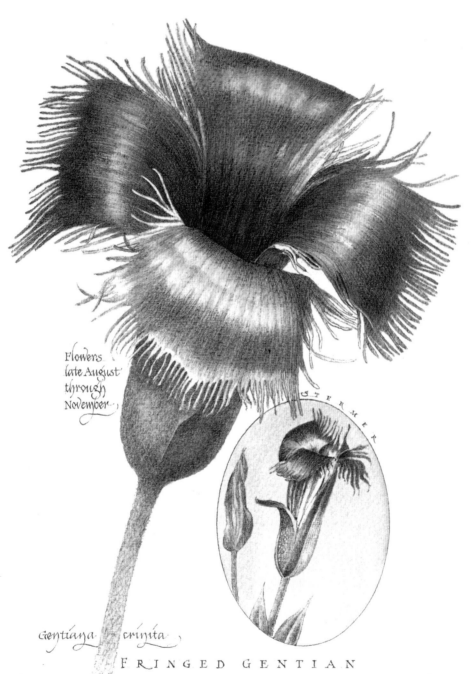

Flowers
late August
through
November

S T E R M E R

Gentiana crinita

F R I N G E D   G E N T I A N

Gladiolus watermeyeri

THIS flowering plant clings
to a precarious existence along the western Cape of
Good Hope around Nieuwoudtville, South Africa. It
is also being cultivated in a California botanical research
and conservation facility, but its future in the wild is
not promising. While the appearance of the flower may
seem unimpressive, its scent is most attractive, similar
to that of violets. The smell is so strong in the field that
the plants can be located by it alone.

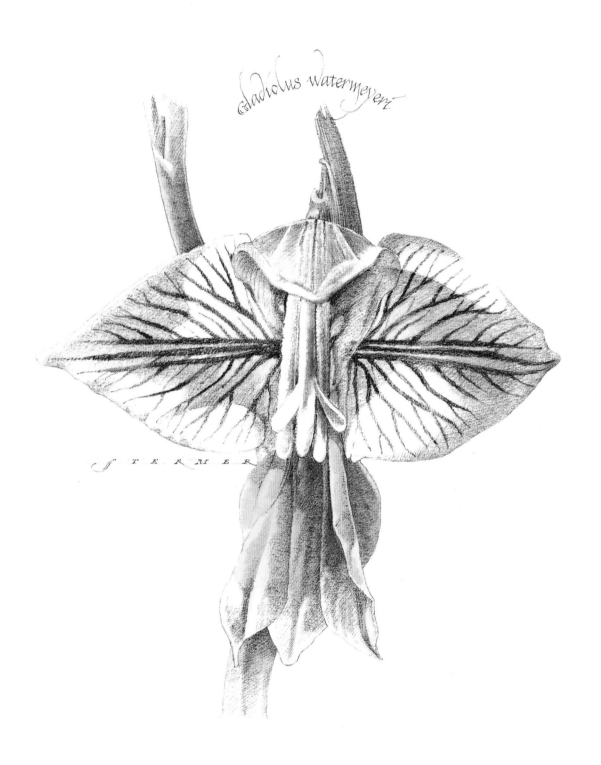

Gladiolus watermeyeri

STERMER

Habenaria leucophaea

PRAIRIE WHITE-FRINGED ORCHID

THIS species has been deemed threatened by the U.S. Fish and Wildlife Service since 1989. It is currently limited to fifty-five sites in seven states and one Canadian province.

H. leucophaea is a close relative of *Platanthera leucophaea*, the eastern prairie fringed orchid, another threatened species. For this reason, it's known to some as *Platanthera praeclara*. The two are distinguished by details of their structure and pollination processes. The western species has larger flowers, and the pollen is collected on the eyes of visiting moths; with the eastern variety, pollen sticks to the moth's proboscis. Both are serviced by night-flying hawk moths.

Habenaria leucophaea

Steiner

PRAIRIE·WHITE-·FRINGED·ORCHID

Hibiscadelphus distans

KAUAI HAU KUAHIWI

ALL ten surviving individuals of this species are found in Waimea Canyon in a state-owned forest reserve on Kauai, Hawaii. It was first discovered in 1972 and was federally listed as endangered fourteen years later.

The rocky bluff where *H. distans* is now found is the remnant of a dryland forest that receives sixty inches of rainfall annually. The area has become degraded and depleted by a large herd of introduced feral goats that destroy ground-covering vegetation and trample and eat seedlings. Other threats to the tiny population include illegal collection, hiking, and wildfire.

Hibiscadelphus distans

Steiner

KAUAI • HAU
KUAHIWI

Hudsonia montana

MOUNTAIN GOLDEN HEATHER

From the mid-1900s to 1978, this plant was considered extinct. Then some specimens were discovered, and it was listed as threatened in 1980. It now exists in only five small populations along the Linville Gorge in northwestern North Carolina.

The most serious threat to its survival appears to be the encroachment of other plants, such as sand myrtle, which shade out the heather. Other threats include trampling by campers and hikers and the occasional extended drought.

Efforts to transplant the species to other locations have been largely unsuccessful, due to the rarity of Chilhowee quartzite soil, which is necessary to its survival and is found only in the Blue Ridge Mountains.

Hudsonia montana

MOUNTAIN
GOLDEN ~ HEATHER

Hymenocallis coronaria

WHITE SPIDER LILY

THE taxonomy of this species suffers from continuing confusion: It is referred to as *H. caroliniana*, *H. crassifolia*, *H. occidentalis*, and *H. coronaria*. This fragile survivor, which lives along the rivers of Alabama, Georgia, and South Carolina, has a grooved stem, which distinguishes it from related species, as well as assisting it in resisting strong currents when the rivers rise. Nevertheless, it is threatened by sewage and other pollutants and is a candidate for protection by the U.S. Fish and Wildlife Service.

WHITE SPIDER LILY

Hymenocallis
coronaria

Stermer

Hymenoxys acaulis var. *glabra*

LAKESIDE DAISY

THIS flowering plant is native to both the United States and Canada: In Ontario, where it is found in twelve sites, it is considered rare; in Ottawa County, Ohio, it is known from only one population, scattered around seven sites, and has been listed as threatened. While it might be assumed that its common name was derived from the plant's predilection for living in proximity to Lake Huron, such is not the case. Rather, it refers to Lakeside, Ohio, near one of its best-known sites.

The state of Ohio has acquired nineteen acres of this daisy's preferred habitats for its protection. Also, the plant is quite easy to grow in cultivation, and attempts are being made to reestablish seedlings in places where it used to thrive.

Hymenoxys acaulis var. glabra

STERMER

L A K E S I D E

D A I S Y

Hyophorbe amaricaulis

M A U R I T I A N   P A L M

T H E tree pictured opposite
is the last remaining individual of its race, a living fossil
destined for extinction. This tree produces flowers and
fruits annually, but none of the fruit have germinated,
or seem able to. Often, the male flowers of a plant will
ripen three or four weeks before the female flowers
become receptive to pollen, which prevents self-pollina-
tion. This may be the case with this palm. There have
been some attempts to pollinate the female flowers, but
so far nothing has worked.

As a final insult, although the plant is as endangered
as any species on the planet, it has yet to be officially
listed as such.

MAURITIAN
PALM

Sterner

Hyophorbe amaricaulis

Iliamna corei

PETER'S MOUNTAIN MALLOW

FROM 1986 through 1989, only three individual plants of this species were found, all from one site on Peter's Mountain in Giles County, western Virginia. Then in 1990, a fourth individual was spotted.

I. corei is a seriously endangered species, as it has been officially termed since 1986. Conservation efforts include fencing in the remaining plants, watering them regularly, and pruning back neighboring plants. Seed collection and germination has also produced a large seed bank under cultivation, greatly improving the species' prognosis.

Iliamna corei

PETERS · MOUNTAIN · MALLOW

Iliamna remota

KANKAKEE MALLOW

THIS is quite similar to
I. corei (see pages 98–99), except in its habitats, growth
habits, and inflorescence structure.

The Kankakee mallow is a native of Langham Island
in the Kankakee River in northern Illinois. In 1834 the
island was the property of Shaw-waw-nas-see, a Pota-
watomie Indian who may have grown the mallow for its
showy flowers or for its leaves, which are reported to be
good to chew. It wasn't classified, however, until the
1870s when the Reverend E. J. Hill came across it while
he was a teacher at Kankakee High School. The species
has been listed as endangered by the state of Illinois
since 1980. A federal listing is currently under review.

Langham Island was incorporated into the Illinois
state park system in the 1940s, and in 1966 it was dedicated as a state nature preserve for the purpose of protecting the mallow from human disturbance.

Thamna remota

K A N K A K E E   M A L L O W

Iris lacustris

DWARF LAKE IRIS

THIS little iris has been
listed as threatened since 1988 and is found at only
sixty sites along the shores of Lake Huron and Lake
Michigan. Despite the flower's status, it is estimated
that only twenty percent of its sites in Michigan and
Wisconsin receive any kind of protection. As the litany
goes, this plant is also threatened by habitat destruction
due to development, especially road building.

Iris lacustris

Sterner

D W A R F   L A K E
I R I S

Isotria medeoloides

S M A L L   W H O R L E D   P O G O N I A

THERE are now approx-
imately fifteen hundred plants of this species remaining
in the wild. They are spread throughout the northeastern United States and on into Ontario, Canada. On
September 10, 1982, *I. medeoloides* was listed as endangered by the U.S. Fish and Wildlife Service.

Several of the sites of the small whorled pogonia, a
terrestrial, woodland orchid, have been destroyed by
residential development, while other populations have
been critically depleted by uninformed collectors.
Recovery strategy includes habitat protection and management and environmental education.

· Isotria medeoloides ·

Flowers: 3/4" long

Leaves: 3" long

Height: 4"~10"

S M A L L · W H O R L E D · P O G O N I A

Sterner

Lavatera assurgentiflora ssp. *glabra*

MALVA ROSA

DESPITE this plant's aggressive, colonizing nature, it is being depleted along its California coastal range because of the grazing of cattle, sheep, and feral goats. As of 1979, it was proposed for addition to the state endangered list. Nevertheless, it is being cultivated successfully as an ornamental in gardens adjacent to its natural habitat.

Lavatera assurgentiflora ssp. glabra

f TERMER

M A L V A

R O S A

Lewisia tweedyi

TWEEDY'S LEWISIA

THIS showy flower, found in rocky areas of the mountains of Washington and British Columbia, is one whose primary threat comes from collectors and amateurs who would rather see it in their rock gardens than flourishing in the wild. Also posing a threat is logging throughout its habitat.

One of its most unusual features is how its seeds are dispersed. *L. tweedyi*'s seeds—actually, attachments to the seeds called elaiosomes—are attractive to ants, who carry them to their nests, where they feast on the elaiosomes, with no harm to the seeds themselves.

Lewisia tweedyi

Stermer

Tweedy's Lewisia

Lilium iridollae

POT-OF-GOLD LILY

·

PANHANDLE LILY

L. IRIDOLLAE is the most recently discovered of the six wild lilies native to eastern North America, and the rarest. In the late 1940s, the foremost amateur lily collector of the time, Mary Gibson Henry, named it after what is found at the end of a rainbow, to commemorate her joy in its discovery. It, too, is a candidate for protection by the federal government. Recent searches for wild populations have been fruitless, and it may be that by the time of this book's publication this lily will have become extinct in the wild.

POT~OF~GOLD
LILY

Sterner

• Lilium iridollae •

Lilium occidentale

WESTERN LILY

THIS extremely rare lily has an amazing history. One of its few sites is on Western Table Bluff in Humboldt County, California. In 1897 Carl Purdy, a collector and seller, described the plant based on specimens taken from that location. After several decades of decimation by collectors, only fifty Western lilies remained on the bluff, announced amateur lily grower A. M. Vollmer. That led to a final—or so it seemed—run on the plant, and by 1940 the population was thought to have been destroyed. During World War II, entry to the location was restricted, and by 1946 a dozen plants flowered; by 1952 more than a hundred plants were located. Back came Vollmer, who, in 1955, publicized to the world of lily pickers that the population had made a comeback to about three hundred plants. As a result, the species was again thought to have been picked into oblivion. Today, the location of the small population that remains is kept a secret, especially from the likes of A. M. Vollmer.

Lilium occidentale

WESTERN LILY

Lilium philadelphicum

WOOD LILY
·
TIGER LILY

Centuries ago Native Americans ate the bulb of this plant, which was found across most of the northern half of the United States. However, since World War II the number of populations has drastically declined, and the remaining sites are kept secret. Nevertheless, the plant is officially considered to be rare only in Colorado. The extraordinary beauty of the flower may be the cause for its decline at the hands of thoughtless but admiring collectors.

Lilium philadelphicum

WOOD LILY or TIGER LILY

Lodoicea maldivica

COCO DE MER

THIS strangely anthropo-
morphic seed, two to three times the size of a coconut,
is the largest in the world. The tree it comes from is
found on two small islands, Praslin and Curieuse, in
the Seychelles group. The total number of trees is
estimated at four thousand.

For several hundred years after the seeds were first
found floating around the Indian Ocean and washed up
on shore, their origin was unknown. Myths sprung up,
one being that they came from under the water, which
led to the tree's common name, French for "sea coconut." Even today, the nuts are thought by some people
in the Seychelles to have aphrodisiac properties.

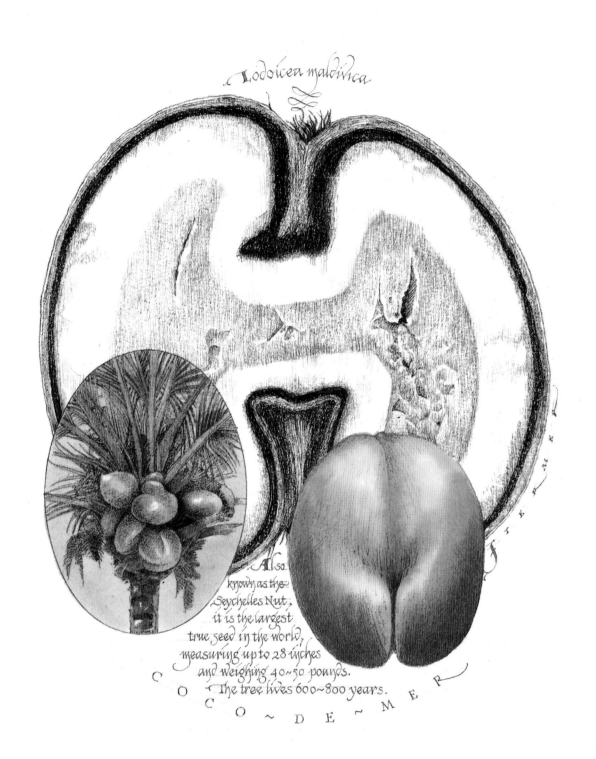

Lodoicea maldivica

Also
known as the
Seychelles Nut,
it is the largest
true seed in the world,
measuring up to 28 inches
and weighing 40~50 pounds.
The tree lives 600~800 years.

COCO~DE~MER

Lupinus aridorum

SCRUB LUPINE

In 1987, the year *L. aridorum* was listed as endangered, it existed at only sixteen sites with a total population of around 350 individuals. Since then the numbers have dwindled.

Its habitat is the dry, sandy soils between Orlando and Walt Disney World, and between Winter Haven and Auburndale, Florida, where large tracts are being converted to housing developments. Florida law restricts the collecting and sale of endangered plants but offers nothing in the way of habitat protection.

Lupinus aridorum

SCRUB · LUPINE

Lysimachia asperulaefolia

ROUGH-LEAVED LOOSESTRIFE

THIS plant, listed as endangered since 1987, is known from only nine populations in North Carolina, the strongest of which is on land owned and managed by The Nature Conservancy.

Interestingly, fire suppression has led to the decline of *L. asperulaefolia*. Without regular burns, its habitat is gradually overtaken by nearby shrubs, placing the loosestrife in shade, which it cannot tolerate. Seven of the nine surviving populations are on land with managed or natural periodic fires, and the plants there are thriving.

Lysimachia asperulaefolia

ROUGH~
LEAVED LOOSESTRIFE

STERMER

Manihot walkerae

WALKER'S MANIOC

LOST for a time, this plant has now been located at just one site in the United States, in Hidalgo County, Texas. It is possible that *M. walkerae* exists in Mexico, but its status there is unknown. It was placed on the endangered list in 1991.

Protection and cultivation attempts are somewhat more intense for this plant than for others because it is closely related to cassava (*Manihot esculenta*), an important food source in the tropics. It is thought that cassava yields may be improved by the addition of genetic material from a wild relative, increasing the crop's resistance to disease or harsh climatic conditions.

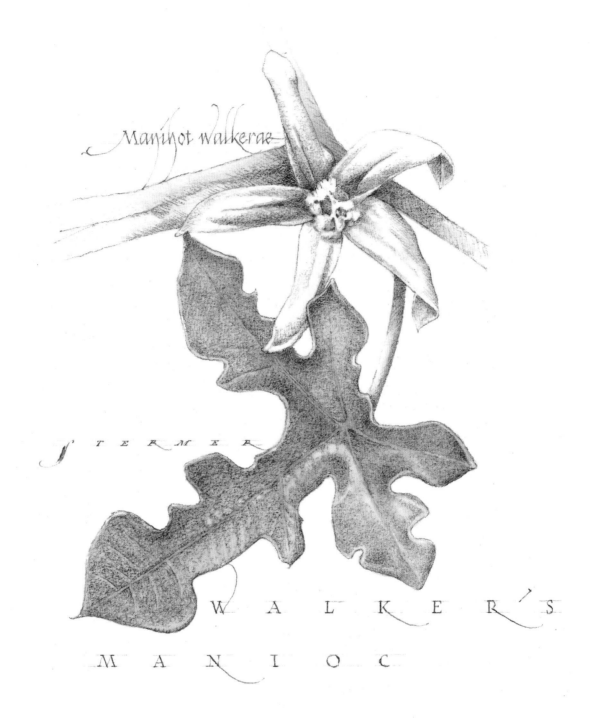

Manihot walkerae

STERMER

WALKER'S

MANIOC

Mimulus glabratus var. *michiganensis*

MICHIGAN MONKEY-FLOWER

THIS species survives at twelve sites in the state for which it is named, two of which contain but two or three individuals. It has been listed as endangered since 1990, but two-thirds of the plants exist on private land, severely limiting what can be done to protect them. Although there doesn't seem to be widespread trade in and collecting of the monkey-flower, there is some; in fact, one of the populations was discovered when a botanist was served his dinner in a restaurant and found a sprig of *M. glabratus* var. *michiganensis* as a garnish on his plate.

Mimulus glabratus var. michiganensis

STAMEN

MICHIGAN
MONKEY~FLOWER

Mirabilis macfarlanei

MACFARLANE'S FOUR-O'CLOCK

THIS perennial, listed as endangered for over fifteen years, was first described in 1936 when Ed MacFarlane, a boatman on the Snake River, pointed it out to two of his passengers, who turned out to be botanists. It is to them that the boatman owes his immortality.

His namesake currently survives in only two populations, comprising seven colonies, spread over sixty acres in Oregon and Idaho. Because MacFarlane's four-o'clock exists in such low numbers, and has such limited distribution, it faces the whole litany of human-induced as well as natural threats. Further, only a small portion of its range exists on public land. It has been listed as endangered since 1979.

Mirabilis macfarlanei

MAP LABELS:
Pacific Ocean
WASHINGTON
Salmon River
Snake River
OREGON
IDAHO
CALIFORNIA
NEVADA

MACFARLANE'S FOUR~O'CLOCK

Stermer

Moraea loubseri

PEACOCK MORAEA

THIS lovely plant is an
example of a relatively new, but growing, category of
life-form; it is extinct in the wild but hangs on to existence in "captivity," thanks to the dedicated efforts of
biologists and botanists.

This species' hero is Harold Koopowitz, who
rescued specimens of the plant from its only habitat in
his native South Africa just before the land was to be
carved up into a stone quarry. Koopowitz sent seeds to
colleagues around the world and brought some to the
University of California at Irvine, where he is the direc-
tor of its arboretum and endangered plant gene bank.
The single mature *M. loubseri* in North America was
brought to flower under his care.

Moraea loubseri

6" ~ 8" ht.

SOUTHERN
AFRICA

Olifants Kop, Langebaan,
native habitat of
Moraea loubseri

CAPE of GOOD HOPE

P E A C O C K · M O R A E A

Sterner

Nypa fruticans

SHUIYE

ALTHOUGH it is known from seven locations in tropical China, there are also reports of this plant in other parts of Asia and as far away as Australia. It is currently listed as vulnerable.

Shuiye is declining largely because it is found to be useful in several ways. The juice of the spadix—or succulent spike—is about fifteen percent sucrose; the seed is edible; and the foliage can be used for construction and basket weaving. Only a few patches of the plant's population are protected in nature reserves, and even there enforcement is inefficient. In this case, survival has to do with bringing the species into commercial cultivation at least as much as with educating the public about conservation.

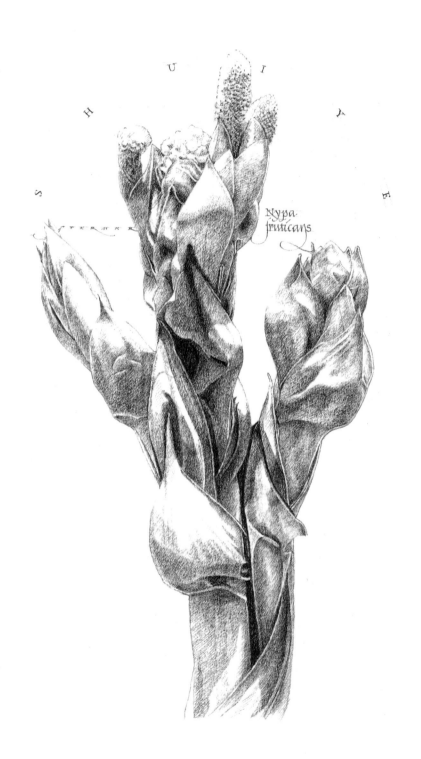

S   H   U   I   Y   E

Nypa fruticans

Oenothera deltoides ssp. *howellii*

ANTIOCH DUNES EVENING PRIMROSE

In 1978 there were approximately one thousand individual plants of this species left in its northern California habitat. Partly because the site was acquired by the U.S. Fish and Wildlife Service, the population soon quadrupled in number. However, it began to decline again, because of heavy industrialization and agriculture adjacent to the area, and it has been on the endangered list since 1993.

Evening primroses have been studied for some time because of their potential medicinal properties. The Antioch Dunes subspecies, in particular, is thought to contain an enzyme, gammalinolenic acid (GLA), that may help control a variety of conditions.

Oenothera deltoides ssp. howellii

ANTIOCH · DUNES · EVENING · PRIMROSE

Orcuttia mucronata

S O L A N O   G R A S S

The north-central California prairie habitat of Solano grass has diminished because of encroaching agriculture, as well as development. Livestock may also play a major part in destroying or impeding the growth of the species. In 1981 The Nature Conservancy removed some grazing horses from a lakebed in the Jepson Prairies Preserve, and within a year, fifty plants were found there. In 1982 fifty-three plants were counted in its only known site; the species had been listed as endangered four years earlier.

Orcuttia mucronata

Sterner

S O L A N O   G R A S S

Paphiopedilum venustum

LADY'S SLIPPER ORCHID

O<small>F</small> the world's twenty thousand orchid species, thirteen hundred live in India —and this is one. They are all threatened with extinction due to massive deforestation for agriculture and other demands brought about by the country's burgeoning population. There is little chance that conservation of any nonedible vegetation will take precedence over the survival of India's people. It appears that the survival of *P. venustum,* as well as the other members of its family, rests on increased knowledge of the species, which could lead to its cultivation and relocation.

LADY'S · SLIPPER · ORCHID

Sterner

· Paphiopedilum venustum ·

Pedicularis furbishiae

FURBISH LOUSEWORT

BELIEVED to have been extinct until 1976, when seven stands of Furbish lousewort were sighted along the St. John River in Maine, the plant was listed as endangered in 1978. By 1985 the total population was estimated at five thousand.

The plant is named for the amateur botanist Kate Furbish, who first described it as a separate species in 1880. The common name—lousewort—comes from the old farmer's notion that the genus harbored lice; "wort" is Anglo-Saxon for plant.

The future of *P. furbishiae* is almost entirely dependent on the success of recovery plans based on maintaining the integrity of its riverbank ecosystem.

Pedicularis furbishiae

Height: 16"~18"

Leaves: 3"~5"l.

FURBISH LOUSEWORT

Steriner

Persea americana

WILD AVOCADO

In one of nature's marvelous symbiotic relationships, the wild avocado—native to Central America, Mexico, and the West Indies—is dependent upon a bird, the rare and beautiful resplendent quetzal (*Pharomachrus mocinno*). Both species are endangered: The avocado tree is used for lumber, and, while the bird is poached to make souvenirs for tourists, its habitat is being destroyed.

The avocado contains everything the quetzal requires. On being swallowed, the fruit is digested, but the inner seed, which is strongly protected, is ultimately regurgitated or passes through the digestive system unharmed, to take root wherever it is dropped. A threat to either species is a threat to both. When both are imperiled from different directions, the danger is imminent.

Persea americana

Pharomacrus mocinno

3 · CORREOS DE GUATEMALA · 3 · TRES CENTAVOS

WILD · AVOCADO

IN · BEAK · OF

· QUETZAL ·

Potentilla robbinsiana

ROBBINS' CINQUEFOIL

·

DWARF CINQUEFOIL

I⊤ isn't always rapacious developers who destroy wild species; friends can do their share of damage. Only one of the four known populations of *P. robbinsiana*—all on Mount Washington in New Hampshire—has survived the plant fanciers and hikers there; it is barely hanging on. Compounding the problems posed by being collected and trod upon, once the cinquefoil is disturbed, it is very slow to recover.

Since the plant's listing as endangered in 1980, a recovery plan was formed with three major objectives: to protect the entire existing population; to establish four new self-sustaining populations within the plant's historic range; and to encourage natural expansion. To those might be added a fourth: to protect it from its friends.

Potentilla robbinsiana

ROBBINS' or DWARF

CINQUEFOIL

Rafflesia microbilora

Upon its discovery in 1983, *R. microbilora* was already endangered: As the flower's size makes it perhaps the largest in the world, the wonder is that it took so long to be noticed by scientists.

All of the twelve *Rafflesia* species are endemic to Malaysia, with this one a native of northern Sumatra. It is a parasite, living entirely within the root tissues of a small variety of jungle vines (*Cissus* ssp.) and thoroughly at their expense. After the flower develops within the root for as long as two years, the basketball-size bud opens with a hiss and then extends to its full size, which may be as large as thirty-eight inches across. After several days, it falls apart and quickly decomposes. When they're blooming, the flowers smell of rotting meat, which attracts pollinating flies, essential for the plant's reproduction.

Large, heavy hoofed animals are just as important: Inside the plant's fleshy fruit are tiny seeds that rely on the hooves of such animals to pick them up and then press them into the ground where they can come into contact with the root system of a new host vine.

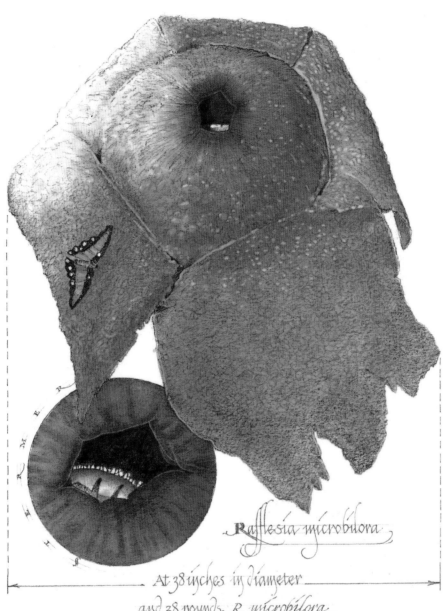

Rafflesia microbilora

At 38 inches in diameter and 38 pounds, *R. microbilora* may be the largest flower on earth.

Ramosmania heterophylla
Randia heterophylla

CAFE MARRON

THIS plant is represented by a single surviving individual on Rodrigues Island, one of a remote group off the coast of Africa that includes Mauritius as well as several smaller islands. Because they are out of normal sailing and trade routes, the islands weren't settled until the seventeenth century, after which human encroachment, along with attendant livestock, quickly began to destroy much of the native flora and fauna. The dodo is Mauritius's most famous example of an endemic species being forced by an introduced species into almost instant extinction.

This plant is thought by Rodriguans to contain medicinal and even religious properties; they will regularly throw coins at its base, asking for grace. Branches and leaves are employed to help the passing of kidney stones and to ward off evil spirits. As long as it is worshiped, it survives.

Ramosmania heterophylla

A · K · A

Randia heterophylla

STERMER

AFRICA

CAFE · MARRON

• Mauritius

Ranunculus cymbalaria

SEASIDE CROWFOOT

BECAUSE this species ranges widely across the eastern United States, it is not federally listed, but it is considered endangered in Wisconsin: In fact, since the wetlands in its last known habitat there were filled, no plants have been seen since 1989.

The seaside crowfoot prefers to live in marshes, muddy banks, shores, ditches, and seepage areas; the plant seems to flourish in brackish waters. That, as well as the fact that it is poisonous, may account for the tendency of livestock to leave it alone.

SEASIDE CROWFOOT

Sternher

• Ranunculus cymbalaria •

Reynoldsia sandwicensis

ʻO H E   M A K A I

This is one of more than twelve hundred threatened plant species native to Hawaii. Development, overcrowding, habitat destruction, and introduced species have wreaked havoc with animal life as well as plants. For example, almost one-third of Hawaii's native birds are believed to be extinct, and more than half of the rest are endangered.

Reynoldsia sandwicensis

'OHE MAKAI

Rhododendron roseum

SWAMP HONEYSUCKLE

·

MOUNTAIN AZALEA

·

ROSE AZALEA

THIS species is found in decreasing numbers along the eastern seaboard from Canada to Virginia, and west to Illinois and Missouri. It is somewhat more common at higher elevations. It has, as yet, not been thought to require federal or state protection.

• Rhododendron roseum •

Flowers: 1½"~2"

2'~10' in height

R. roseum is known by various
names in various locales:
Rose Azalea, Mountain Azalea,
Woolly Azalea & Early Azalea,
among others.

Leaves: 2"~4"

S W A M P   H O N E Y S U C K L E

Stenner

Romulea

THE genus *Romulea* refers to
a number of species mainly confined to Africa and the
countries bordering the Mediterranean Sea. The flower
pictured is quite rare, and its habitat is shrinking.

STERMER.

R O M U L E A *sp.*

Ruellia humilis

WILD PETUNIA

MUCH of this plant's range—Wisconsin south to Texas, Pennsylvania west to Nebraska—is in prairie habitat, which is itself also endangered. For each bushel of corn harvested in Wisconsin, two bushels of soil are washed down the Mississippi. To compound the problem, the plants and animals that maintain the fragile formula for prairie soil in their genes have also not been protected.

WILD ✦ PETUNIA

Zermer

✦ Ruellia humilis ✦

Sarracenia oreophila

GREEN PITCHER PLANT

ABOUT twenty-six col-
onies, ranging in size from one plant to over one thousand, currently survive in Alabama and Georgia. *S.
oreophila* has been listed as endangered throughout its
range since 1979.

One of the threats to its existence, apart from habitat disruption, is illegal horticultural trade in wild
plants. Like the Venus flytrap (see pages 62–65), it is
much sought after because it is an insectivore. Insects
are first attracted either by nectar secreted by the plant
near the top of its pitcher, or by its color. When an
insect crawls into the pitcher, it is trapped and even-
tually digested by plant enzymes.

No pitcher plants—in fact, no rare carnivorous
plants—should ever be taken from the wild. Rather, the
fancier should purchase plants from only those reputable experts who cultivate their own.

Sarracenia oreophila

S T E R M E R

G R E E N

P I T C H E R

P L A N T

Silene petersonii var. *petersonii*

PLATEAU CATCHFLY

THIS plant, along with a closely related cousin, *S. petersonii* var. *minor*, is a candidate for listing as endangered and is among the 114 vulnerable plants in Utah, eighty-five of which are found in no other state. Although the plateau catchfly grows at elevations of eleven thousand feet, even there it is threatened by oil and gas exploration, livestock grazing, and recreation.

Silene petersonii var. petersonii

STERMER

PLATEAU

CATCHFLY

Sphaeralcea caespitosa

JONES' GLOBE MALLOW

IN an apparent contradiction, this plant is both abundant and rare; abundant where it grows, but it survives only in two counties in Utah and in one in Nevada. It is a candidate for listing as endangered or threatened, because changing land use, common in that area, could wipe the species out in short order. Conservation efforts in the two states are among the least developed of any in North America. Neither state has a natural heritage program, and as yet there is insufficient support for significant efforts on the part of The Nature Conservancy.

Sphaeralcea caespitosa

J O N E S'
G L O B E
M A L L O W

SPIDER ORCHID

Th is officially undescribed (no Latin name) spider orchid is known from only one plant on Mount Canobolas, near Orange, in southeastern Australia. Because of its unusual environment, as well as its isolation from the earth's other great landmasses, Australia sustains many unusual flora, eighty percent of which are endemic. The major threats to Australia's natural habitats stem from two hundred years of European settlement, much the same as North America, South America, and elsewhere.

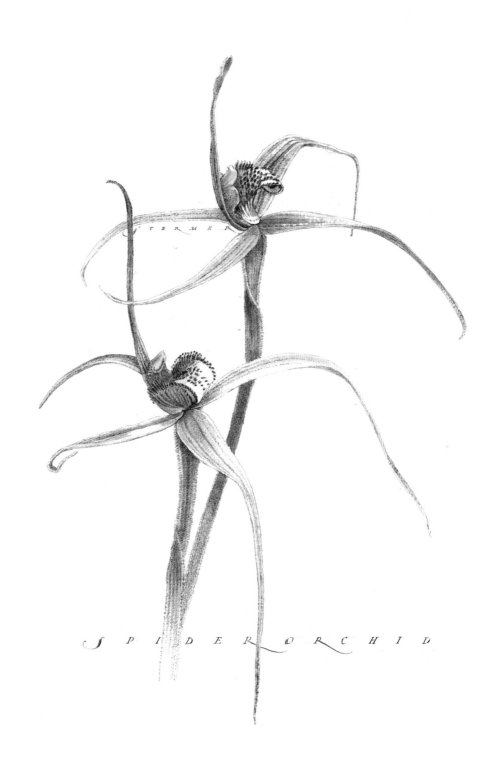

TERMER

SPIDER ORCHID

Stewartia malacodendron

WILD CAMELLIA

·

SILKY CAMELLIA

THE Garden Club of America considers this gorgeous flower endangered, even if the U.S. Fish and Wildlife Service does not as yet. Its range is primarily in the Carolinas, although it is also found in Virginia, Georgia, Florida, east Texas, Alabama, Mississippi, Tennessee, and Arkansas.

Stewartia malacodendron

WILD or
SILKY
CAMELLIA

Styrax texana

TEXAS SNOWBELLS

THIS plant, listed as endangered since 1984, exists only in about forty shrubs in south-central Texas. Its low numbers, compounded by the apparent failure of its reproductive system, make the population particularly vulnerable to extinction.

Texas snowbells were brought to this precarious situation by grazing cattle, deer, and imported sheep, along with stream-bank erosion due to flash flooding. The U.S. Fish and Wildlife Service is augmenting traditional conservation efforts on behalf of this species, with attempts to transplant seedlings into greenhouses, there to be raised and then reintroduced into safer sites.

Styrax
texana

TEXAS

SNOWBELLS

Tecophilaea cyanocrocus

AZULILLO

THIS native of Chile is now considered extinct in the wild, primarily due to urban expansion and development, but also because of collection; it is the only known member of its botanical family to have edible bulbs. Both its species designation and its common name refer to the flower's unusually bright blue color.

Tecophilaea cyanocrocus

Trifolium stoloniferum

RUNNING BUFFALO CLOVER

By 1984 this plant was known to exist in only two populations, both in West Virginia. After it was finally listed as endangered in 1987, several other sites were found in Indiana, Ohio, and Kentucky.

Botanists have speculated that the clearings preferred by this plant were maintained by grazing buffalo herds that migrated along its original range. The herds have, of course, disappeared; along with them have gone the clearings and most of the running buffalo clover.

Conservation efforts are under way and include the protection of existing sites, along with transplantation efforts to establish new, less vulnerable, populations.

Trifolium stoloniferum

STERMER

RUNNING

BUFFALO

CLOVER

Trillium persistens

PERSISTENT TRILLIUM

T. PERSISTENS was first described in 1971, and it soon became one of the first federally listed endangered plant species. In 1979 it was featured on a United States commemorative stamp in the Endangered Flora set. The name "persistent" was chosen because the plants emerge early in the spring (March or even February) and remain green above ground until September or October, after the stems and leaves of other trilliums have withered.

Persistent trillium exists only in Tallulah Gorge in northeastern Georgia and South Carolina, site of the nine-hundred-foot gulf that "The Flying" Karl Wallenda tightroped across in 1970.

Trillium persistens

SUMMER

PERSISTENT TRILLIUM

Trollius laxus ssp. *laxus*

SPREADING GLOBEFLOWER

The status of this rare plant is currently under review for federal listing and protection. It is found, if at all, in Connecticut, New Jersey, New York, Ohio, and Pennsylvania. It has suffered because of the destruction of its wetland habitats and because of the introduction of another species, purple loosestrife (*Lythrum salicaria*), which takes over wet areas, pushing out many kinds of native vegetation.

Trollius laxus
ssp. laxus

2" W.

2' h.

S P R E A D I N G   G L O B E F L O W E R

Steiner

Vicia menziesii

HAWAIIAN VETCH

IN 1978 Hawaiian vetch had the distinction of being the first plant native to Hawaii to be listed as endangered by the U.S. Fish and Wildlife Service. It had been considered extinct until it was rediscovered in 1974 on the slopes of Mauna Loa.

When Captain Cook arrived, the flora and fauna of the Hawaiian Islands had evolved for centuries, uninterrupted by foreign factors. However, since the islands' discovery by Europeans (and colonization by Americans), introduced livestock, land development, agriculture, urban sprawl, pollution, and waves of human invaders have seriously altered their ecology and endangered nearly half the native species of both plants and animals.

Vicia menziesii

Flowers: 1¼" l.
Pods: 4" l ~ ¾" w

HAWAII

N

Mauna Kea

Mauna Loa

V. MENZIESII
Former Habitat
Present Habitat

H A W A I I A N   V E T C H

Sternmer

Welwitschia mirabilis

THIS is the most interesting and renowned plant of the Namib Desert, which is located on the coast of southwestern Africa. The species has been protected since the beginning of the twentieth century, but only a few specimens survive.

W. mirabilis is of special interest for several reasons. One is that it is unusually long-lived; a particularly large individual has been estimated as being over five hundred years old. Another feature is that it is adaptable to extremes of environmental conditions. Finally, it exists between two botanical groupings, scientifically isolated in the plant kingdom.

The foot-high trunk is covered with cork, which soaks up water. The deep root draws water from the sand, and a side root system takes up any other moisture that the Namib's rare rains provide.

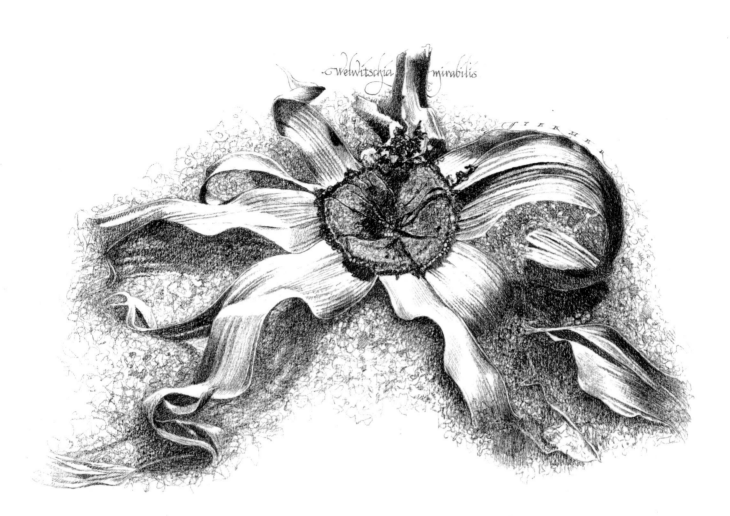

ACKNOWLEDGMENTS

I required the assistance of many people during the course of working on this book. One was indispensable: Megan Stermer-Gregory was a dogged compiler, researcher, editor, and cattle prod. Without my daughter's contribution, along with the expert research of Shellei Addison and Jill Westley, the masterly editing of Sharon AvRutick, and the elegant craft of Sam Antupit, this would not have come together.

Over a decade ago, Barbara Cady, the editor and publisher of *Flowers&* magazine, commissioned me to produce my first twelve pictures of endangered plants, an act of reckless faith. Soon after, Fran Gendlin, then editor of *Sierra* magazine, and Bill Blair of *Country Journal* each published several illustrations in the growing series. Peter Mayer, owner and publisher of The Overlook Press, also offered early support for this project. To each of them I owe a debt.

This is a book by a layman for laypeople. The following experts were both unfailingly generous with their knowledge and inspiring in their commitment to the cause of conservation: Ken Berg, Catherine Caulfield; R. A. DeFilipps, Office of Biological Conservation, Smithsonian Institution; the Departments of Natural Resources in the fifty states; June Dobberpuhl, State of Wisconsin Department of Natural Resources; John Garst; Derral Herbst, botanist, Office of Environmental Services, Honolulu; Fred Hill; Janet Hohn; Alice Q. Howard, National Alliance for Plants, Oakland, California; Alison Jolly; Harold Koopowitz; Mrs. Lothian Lynas, New York Botanical Garden; Charlie McDonald, Fish and Wildlife Service, Office of Ecological Services for region 2; Linda McMahon; Ralph Martin, Department of Biology, Brooklyn College; Tom Morley; all state and local native plant societies; The Nature Conservancy; A. W. Owadally, Conservator of Forests, Mauritius; Tom Patrick, Georgia Department of Natural Resources, Wildlife Resources Division; Mrs. Edward King Poor III, Garden Club of America; Ghillean T. Prance, Kew Gardens, London; John Schwegman, Illinois Department of Conservation; Paul Somers, Tennessee Department of Conservation; Tennessee Native Plant Society; United States Department of the Interior, Fish and Wildlife Service; R. E. Vaughan, Mauritius; and Paul Wiegman.

It would be impossible to attempt a work of this kind without leaning heavily on an army of botanists, photographers, conservationists, and other field workers. I relied specifically on the following, and the fact that they were unaware of their contribution makes them no less appreciated.

Richard M. Adams II, University of California Botanical Gardens at Riverside; L. H. Bailey; Colin Bower; Louise K. Broman; Gerald Carr, Department of Botany, University of Hawaii; Frederick W. Case II; Ann S. Causey, Department of Botany, Auburn University; Beecher Crampton, Department of Agronomy and Range Science, University of California, Davis; Mark Dimmitt, Arizona-Sonora Desert Museum, Tucson; Richard Dyer; Donald C. Eastman; L. M. Eastman; O. L. Eye; John Farrar; Michael and Patricia Fogden; Kerry T. Givens; the Iris Hardwick Library of Photographs; Ann Howald, California Native Plant Society; William E. Jennings; Charles Johnson; S. Junak; Douglas Kirkland (SYGMA); Kitty Kohout; Don Kurz; the library of the New York Botanical Garden, Bronx; Phillip Malnassy; the Michigan Natural Features Inventory; Robert W. Mitchell; Paul M. Montgomery; Helen Margaret Mulligan, Seattle Garden Club; Arnold Newman; the Ohio Department of Natural Resources; the Photographic Service Corporation, Springfield, Illinois; Martin W. Poulson; Noble Proctor; Werner Rauh, University of Heidelberg Institute of Systematic Botany; Robert Read; Andy Robinson; J. H. Robinson; Alfred Schotz; John Shaw; James Shevock, United States Forest Service, Sequoia National Park; Rob Simpson; the Harry Smith Horticultural Photographic Collection; LaVerne Smith; Tom Smith; Hugh Spencer; Myrna P. Steinkamp; W. Vent, Berlin; Eric Vlaszek, Darien, Illinois; Kerry S. Walter; Michael Warren; Virginia Weinland; Michael Willets; Joyce R. Wilson; and Kenneth Wurdack.

A P P E N D I X

The following is a partial listing of the many organizations and institutions involved in the protection and recovery of rare and endangered flora. It is presented here with the hope that you will find a way to help that matches your concerns and abilities.

Assistance is welcomed on many levels, from membership in one or more of the national and international conservation organizations, such as the World Wildlife Fund or The Nature Conservancy, to more direct involvement through any of the native plant societies, natural heritage programs, or botanical societies; these last can be found in your telephone book or by calling the botanical department at a local college or university.

UNITED STATES AND CANADA

Abundant Life Seed Foundation
Box 772, 1029 Lawrence
Gardiner, Washington 98334

Alabama Wildflower Society
Route 2, Box 410
Northport, Alabama 35476

Alaska Conservation Society
Box 80192
College Branch
Fairbanks, Alaska 99708

American Conservation Association, Inc.
30 Rockefeller Plaza, Room 5425
New York, New York 10020

American Horticultural Society, Inc.
Mount Vernon, Virginia 22121

The American Museum of Natural History
Central Park West at 79th Street
New York, New York 10024

American Nature Study Society
R.D. 1
Homer, New York 13077

The Arctic Institute of North America
University Library Tower
2920 24th Avenue, N.W.
Calgary, Alberta T2N 1N4
Canada

The Arizona Native Plant Society
P.O. Box 41206 Sun Station
Tucson, Arizona 85717

Arizona Natural Heritage Program
30 North Tucson Boulevard
Tucson, Arizona 85716

Association of Western Native Plant
   Societies
4949 N.E. 34th
Portland, Oregon 97211

Atlanta Botanical Garden
P.O. Box 77246
Atlanta, Georgia 30357

The Atlantic Center for the Environment
951 Highland Street
Ipswich, Massachusetts 01938

The Berry Botanical Garden
11505 S.W. Summerville
Portland, Oregon 97219

Biodiversity Resource Center
California Academy of Sciences
Golden Gate Park
San Francisco, California 94118

Biological Institute of Tropical America
P.O. Box 2585
Menlo Park, California 94025

Botanical Garden
Centennial
University of California
Berkeley, California 94720

California Botanical Society
Department of Botany
University of California
Berkeley, California 94720

California Native Plant Society
2380-D Ellsworth
Berkeley, California 94704

California Nongame Heritage Program
1416 9th Street, 12th Floor
Sacramento, California 95814

Canadian Nature Federation
Suite 203, 75 Albert Street
Ottawa, Ontario K1P 6G1
Canada

Canadian Wildlife Service
Room 230, 4999-98
Edmonton, Alberta T6B 2X3
Canada

Center for Action on Endangered Species
175 West Main Street
Ayer, Massachusetts 01432

Center for Environmental Education, Inc.
2100 M Street, N.W.
Washington, D.C. 20037

Center for Natural Areas
1525 New Hampshire Avenue, N.W.
Washington, D.C. 20036

Chesapeake Bay Center for Environmental
   Studies
R.R. 4, Box 662
Edgewater, Maryland 21037

Chihuahuan Desert Research Institute
P.O. Box 1334
Alpine, Texas 79830

Connecticut River Watershed Council, Inc.
125 Combs Road
Easthampton, Massachusetts 01027

The Conservation Agency
97B Howland Avenue
Jamestown, Rhode Island 02835

The Conservation & Research Foundation,
   Inc.
Connecticut College
New London, Connecticut 06320

The Conservation Foundation
1015 18th Street, N.W.
Washington, D.C. 20036

Conservation International
1015 18th Street, N.W., Suite 1000
Washington, D.C. 20036

Defenders of Wildlife
1244 19th Street, N.W.
Washington, D.C. 20036

Earthforce Environmental Society
2623 West 4th Avenue
Vancouver, British Columbia V6K 1T8
Canada

Ecological Services Field Office
3616 West Thomas Road, Suite 6
Phoenix, Arizona 85019

Ecological Society of America
Department of Botany
Duke University
Durham, North Carolina 27706

Environmental Defense Fund
1616 P Street, N.W., Suite 150
Washington, D.C. 20036

Envirosouth, Inc.
P.O. Box 1711
Montgomery, Alabama 36117

Fairchild Tropical Garden
11935 Old Cutler Road
Miami, Florida 33156

Friends of Africa in America
330 South Broadway
Tarrytown, New York 10591

Friends of the Earth
124 Spear Street
San Francisco, California 94105

Friends of the Earth (International)
530 Seventh Street, S.E.
Washington, D.C. 20003

Garden Club of America
598 Madison Avenue
New York, New York 10022

Georgia Department of Natural Resources
Freshwater Wetlands and Heritage
   Inventory
Route 2, Box 119-D
Social Circle, Georgia 30279

Global Tomorrow Coalition
1325 G Street, N.W., Suite 915
Washington, D.C. 20036

Greenpeace U.S.A.
1436 U Street, N.W.
Washington, D.C. 20009

Growers Network
RFD 2
Princeton, Missouri 64673

Hawaii Department of Land and
   Natural Resources
54 South High Street
Wailuku, Hawaii 96793

Hawaiian Botanical Society
Botany Department
University of Hawaii
3190 Maile Way
Honolulu, Hawaii 96822

International Institute for Environment
   and Development (IIED)
1717 Massachusetts Avenue, N.W.
Washington, D.C. 20036

Island Resources Foundation
P.O Box 4187
St. Thomas, Virgin Islands 00801

Izaak Walton League of America
1800 North Kent Street, Suite 806
Arlington, Virginia 22209

John Muir Institute for Environmental
   Studies
743 Wilson Street
Napa, California 94558

Louisiana Natural Heritage Program
P.O. Box 98000
Baton Rouge, Louisiana 70898-9000

Maryland Natural Heritage Program
Tawes State Office Building, B-2
Annapolis, Maryland 21401

Minnesota Natural Heritage Program
P.O. Box 7, DNR Building
500 Lafayette
St. Paul, Minnesota 55155

Missouri Botanical Garden
P.O. Box 299
St. Louis, Missouri 63166

Missouri Conservation Department
2901 West Truman Boulevard
Jefferson City, Missouri 65109

Monitor
1522 Connecticut Avenue, N.W.
Washington, D.C. 20036

National Audubon Society
666 Pennsylvania Avenue, S.E.
Washington, D.C. 20003

National Geographic Society
17th and M Streets, N.W.
Washington, D.C. 20036

National Parks and Conservation
   Association
1701 18th Street, N.W.
Washington, D.C. 20009

National Wetlands Technical Council
Suite 300
1717 Massachusetts Avenue, N.W.
Washington, D.C. 20036

National Wildlife Federation
1400 16th Street, N.W.
Washington, D.C. 20036

The Natural Heritage Division
Fish and Wildlife Service
Department of the Interior
18th and C Streets, N.W.
Mail Stop 725
Arlington Square Building
Washington, D.C. 20240

Natural Resources Defense Council, Inc.
122 East 42nd Street
New York, New York 10017

Natural Resources Office
Naval Air Station, North Island (Bldg. 3)
San Diego, California 92135-5018

The Nature Conservancy
1815 North Lynn Street
Arlington, Virginia 22209

Nevada Natural Heritage Program
201 South Fall
Carson City, Nevada 89710

New England Wildflower Society
"Garden in the Woods"
Hemenway Road
Framingham, Massachusetts 01701

New York State Department of
   Environmental Conservation
50 Wolf Road
Albany, New York 12233

North Carolina Natural Heritage Program
P.O. Box 27687
Raleigh, North Carolina 27611

Rachel Carson Trust for the Living
   Environment
8940 Jones Mill Road
Washington, D.C. 20015

Rainforest Action Network
450 Sansome Street
San Francisco, California 94111

Rainforest Alliance
295 Madison Avenue, Suite 1804
New York, New York 10017

The Rare and Endangered Native
   Plant Exchange
c/o New York Botanical Garden
Bronx, New York 10458

River Conservation Fund
317 Pennsylvania Avenue, S.E.
Washington, D.C. 20003

Rocky Mountain Center on Environment
1115 Grant Street
Denver, Colorado 80203

Save the Dunes Council
P.O. Box 114
Beverly Shores, Indiana 46301

Sierra Club
730 Polk Street
San Francisco, California 94109

Smithsonian Institution
Office of Exhibits
10th and Constitution, N.W.
Washington, D.C. 20560

South Carolina Heritage Trust Program
P.O. Box 167
Columbia, South Carolina 29202

Tallgrass Prairie Foundation
5450 Buena Vista
Shawnee Mission, Kansas 66205

Tennessee Department of Conservation
Ecological Services
701 Broadway
Nashville, Tennessee 37219-5237

Tropical Ecosystem Research
   and Rescue Alliance
Terra International
Washington, D.C. 20036

United Nations Environment Programme
UNEP Liaison Office
Ground Floor
1889 F Street, N.W.
Washington, D.C. 20006

Utah Natural Heritage Program
1636 West North Temple, Suite 316
Salt Lake City, Utah 84116-3193

Wetlands for Wildlife, Inc.
P.O. Box 147
Mayville, Wisconsin 53050

The Wilderness Society
1901 Pennsylvania Avenue, N.W.
Washington, D.C. 20006

Wildlife Preservation Trust
   International, Inc.
34th Street and Girard Avenue
Philadelphia, Pennsylvania 19104

The Wildlife Society, Inc.
7101 Wisconsin Avenue, N.W. #611
Washington, D.C. 20014

World Resources Institute
1709 New York Avenue, N.W.
Washington, D.C. 20006

World Wildlife Fund (WWF)—Canada
60 St. Clair Avenue East, Suite 201
Toronto, Ontario M4T 1N5
Canada

WWF—United States
1250 24th Street, N.W.
Washington, D.C. 20037

Worldwatch Institute
1776 Massachusetts Avenue, N.W.
Washington, D.C. 20036

The National Wildlife Federation publishes
Conservation Directory, an annual listing of
conservation groups and government
agencies.

INTERNATIONAL

Action for Environment
P.O. Box 10-030
Wellington
New Zealand

The Ark Trust
500 Harrow Road
London W9 3QA
England

Australian Conservation Foundation
672b Glenferrie Road
Hawthorn Vic 3122
Australia

Campaign to Save Native Forests
794 Hay Street
Perth W.A. 6000
Australia

Compañía Nacional de Reforestación
(CONARE)
Apartado Postal 17015
El Conde
Caracas 1010A
Venezuela

Conservation Council of Victoria
247 Flinders Lane
Melbourne, Victoria 3000
Australia

The Conservation Foundation
1 Kensington Gore
London SW7 2AR
England

Conservation New Zealand
P.O. Box 12.200
Wellington
New Zealand

Department of Conservation
P.O. Box 10-420
Wellington
New Zealand

Department of Scientific and Industrial
Research, Botany Division
Private Bag
Christchurch
New Zealand

Earthwatch (Europe)
P.O. Box 392
Headington
Oxford OX3 OUE
England

Environment & Conservation Organization
of New Zealand, Inc.
P.O. Box 11.057
Wellington
New Zealand

Environmental Liaison Center
P.O. Box 72461
Nairobi
Kenya

The Flora and Fauna Preservation Society
c/o Zoological Society of London
Regents Park
London NW1 4RY
England

Forest Research Institute
P.O. Box 201
52109 Kuala Lumpur
Malaysia

Friends of the Earth (NZ)
P.O. Box 39-065
Auckland West
New Zealand

Friends of the Earth (UK)
26-28 Underwood Street
London, N1 7JQ
England

Gaia Foundation
18 Well Walk
London NW3 1LD
England

The Garden History Society
12 Charlburg Road
Oxford OX2 6UT
England

Greening Australia
Block D, Acton House
Corner Marcus Clarke Street and
Edinburgh Avenue
Acton Act 2600
Australia

International Dendrology Society
Whistley Green Farmhouse
Hurst, Reading
Berkshire RG10 ODU
England

The International Organization for
Succulent Plant Study
Royal Botanic Gardens
Kew, Richmond
Surrey TW9 3AB
England

The International Society for
Horticultural Science
Ministry of Agriculture
Bezuidenhoutseweg 73
The Hague
Netherlands

International Union for Conservation of
Nature and Natural Resources (IUCN)
Avenue du Mont-Blanc
1196 Gland
Switzerland

Living Earth
10 Upper Grosvenor Street
London W1X 9PA
England

Men of the Trees
New South Wales
11 Pebbly Hill Road
Cattai NSW 2756
Australia

Nature Conservancy Council
North West (Scotland) Region
Old Bank Road
Golspie KW10 6RS
Scotland

Royal Botanic Gardens
Kew, Richmond
Surrey TW9 3AB
England

The Royal Horticultural Society
Vincent Square
Westminster SW1
London
England

Smithsonian Tropical Research Institute
P.O. Box 2072
Balboa, Canal Zone
Panama

Survival International
310 Edgware Road
London W2 1DY
England

United Nations Environment
  Programme (UNEP)
Headquarters
P.O. Box 30552
Nairobi
Kenya

The Wilderness Society
130 Davey Street
Hobart TAS 7000
Tasmania

WWF—Australia
Level 17
St. Martins Tower
31 Market Street
Sydney NSW 2000
Australia

WWF—Malaysia
P.O. Box 10769
50724 Kuala Lumpur
Malaysia

WWF—New Zealand
35 Taranaki Street
P.O. Box 6237
Wellington
New Zealand

WWF—South Africa
P.O. Box 456
Stennenbosch 7600
South Africa

WWF—United Kingdom
Panda House
Weyside Park, Godalming
Surrey GU7 1XR
England

BIBLIOGRAPHY

BOOKS

Abbey, Edward, and the Editors of TIME-LIFE Books. *Cactus Country*. Alexandria, Vir.: Time-Life Books, 1973.

Ahmadjian, Vernon. *Flowering Plants of Massachusetts*. Amherst, Mass.: University of Massachusetts Press, 1979.

Ayensu, Edward S. *Jungles*. New York: Crown Publishers, 1980.

Bailey, Liberty Hyde, and Ethel Joe Bailey, comps. *Hortus Third, A Concise Dictionary of Plants Cultivated in the United States and Canada*. New York: MacMillan Publishing Company, 1976. Revised and expanded by the staff of the Liberty Hyde Bailey Hortorium.

Balouet, Jean-Christophe. *Extinct Species of the World*. New York: Barron's, 1990.

Benoit, I. L., ed. *Red Book of Chilean Terrestrial Fauna, Part One*. Santiago, Chile: CONAF, 1989.

Benson, Lyman. *The Cacti of the United States and Canada*. Stanford, Cal.: Stanford University Press, 1982.

Bianchini, Francesco, and Francesco Corbetta. *The Complete Book of Fruits and Vegetables*. New York: Crown Publishers, 1976.

Borland, Hal. *A Countryman's Flowers*. New York: Alfred A. Knopf, 1981.

Brynildson, Inga. *Wisconsin's Endangered Flora*. Department of Natural Resources, Office of Endangered & Nongame Species.

Burn, Barbara. *North American Wildflowers*. New York: Bonanza Books, 1984.

California Department of Education. *California Endangered Species Resource Guide*. Sacramento, 1993.

Colorado Native Plant Society. *Rare Plants of Colorado*. Estes Park, Col.: the society in cooperation with Rocky Mountain Nature Association, 1989.

Dejey, M. A. *Health Plants of the World*. New York: Newsweek Books, 1975. Adapted from the Italian of Francesco Bianchini and Francesco Corbetta.

de Wit, H. C. D. *Plants of the World, The Higher Plants II*. New York: E. F. Dutton & Co., 1967.

Durrell, Gerald. *Golden Bats and Pink Pigeons*. New York: Simon and Schuster, 1977.

Eastman, Donald C. *Rare and Endangered Plants of Oregon*. Wilsonville, Ore.: Beautiful America Publishing Co., 1990.

Gleason, Henry A., Ph.D. *The New Britton and Brown Illustrated Flora of the Northeastern United States and Adjacent Canada*. New York and London: Hafner Publishing Co., for the New York Botanical Garden, 1963.

Grosvenor, Gilbert, LL.D., Litt.D., D.Sc., and Alexander Wetmore, Ph.D., D.Sc., eds. *The Book of Birds*. Washington, D.C.: National Geographic Society, 1937.

Halliday, Tim. *Vanishing Birds, Their Natural History and Conservation*. New York: Holt Rinehart and Winston, 1978.

Heywood, V. H., and S. F. Chant, eds. *Popular Encyclopedia of Plants*, Cambridge, England: Cambridge University Press, 1982.

Höhn, Reinhardt. *Curiosities of the Plant Kingdom*. New York: Universe Books, 1980.

House, Homer D. *Wild Flowers*. New York: The MacMillan Co., 1935.

IUCN/UNEP. *The IUCN Directory of Afrotropical Protected Areas*. Gland, Switzerland and Cambridge, England: 1987.

Kastner, Joseph. *A Species of Eternity*. New York: Alfred A. Knopf, 1977.

Kennedy, Michael. *Australia's Endangered Species*. Sydney, Australia: Simon & Schuster, 1990.

Kimura, Bert Y., and Kenneth M. Nagata. *Hawaii's Vanishing Flora*. Honolulu: The Oriental Publishing Co., 1980.

Koopowitz, Harold, and Hilary Kaye. *Plant Extinction, A Global Crisis*. Washington, D.C.: Stone Wall Press, Inc., 1983.

Kramer, Jack. *Orchids, Flowers of Romance and Mystery*. New York: Harry N. Abrams, 1979.

Lemmon, Robert S., and Charles C. Johnson. *Wildflowers of North America*. Garden City, N.Y.: Hanover House, 1961.

Lewis, G. Joyce, and A. Amelia Obermeyer. *Gladiolus. A Revision of the South African Species. Journal of South African Botany, Supplementary Volume No. 10*. Cape Town, Johannesburg, London, New York: Purnell, 1972.

Li-kuo, Fu, ed. *China Plant Red Data Book, Rare and Endangered Plants, Volume 1*. Beijing and New York: Science Press, 1992.

Line, Les. *The Audubon Society Book of Wildflowers.* New York: Harry N. Abrams, 1978.

Lowe, David W., John R. Matthews, and Charles J. Moseley, eds. *The Official World Wildlife Fund Guide to Endangered Species of North America, Volumes I and II.* Washington, D.C.: Beacham Publishing, 1990.

Lucas, Gren, and Hugh Synge, comps. *The IUCN Plant Red Data Book.* Morges, Switzerland: IUCN, 1978.

Marinuzzi, Anna-Sofia. *Marvellous World of Exotic Flowers.* London: Abbey Library, 1977.

Mohlenbrock, Robert H. *Where Have All The Wildflowers Gone?* New York: MacMillan Publishing Co., 1983.

Newman, Arnold. *Tropical Rainforest.* New York, Oxford: Facts on File, 1990.

Niering, William A., and Nancy C. Olmstead. *The Audubon Society Field Guide to North American Wildflowers.* New York: Alfred A. Knopf, 1979.

Peterson, Roger Tory, and Margaret McKenny. *A Field Guide to Wildflowers of Northeastern and Northcentral North America.* Boston: Houghton Mifflin, 1968.

Radford, Alfred D., Harry E. Ahles, and C. Ritchie Bell. *Manual of Vascular Flora of the Carolinas.* Chapel Hill, N.C.: University of North Carolina Press, 1968.

Silcock, Lisa, ed. *The Rainforests, A Celebration.* San Francisco: Chronicle Books, 1989. Compiled by the Living Earth Foundation.

Stevens, George T., M.D. *Illustrations of Flowering Plants of the Middle Atlantic and New England States.* Flatbush Printing Co., 1930.

Tosco, Uberto. *The World of Wildflowers and Trees.* New York: Bounty Books, 1972.

U.S. Government Printing Office. "Endangered and Threatened Wildlife and Plants." Washington, D.C., 50CFR 17.11 & 17.12, August 29, 1992.

Ward, Daniel B., ed. *Rare and Endangered Biota of Florida, Volume Five: Plants.* Gainesville, Fla.: University Presses of Florida, 1978–82.

Washington Natural Heritage Program. "An Illustrated Guide to the Endangered, Threatened and Sensitive Vascular Plants of Washington." 1981.

Went, Frits, and the Editors of LIFE. *The Plants.* New York: Time, 1963.

Wilson, E. O., ed. *Biodiversity.* Washington, D.C.: National Academy Press, 1988.

PERIODICALS

American Rhododendron Society. *Quarterly Bulletin.* 32(3), summer 1978.

Arnold Arboretum. *Bulletin of Popular Information.* Series 3, 1(7). Harvard University, May 20, 1927.

Aubeeluck, P., and Y. Boodoo, et al. "State of the environment in Mauritius: a report prepared for presentation at the United Nations Conference on Environment and Development, Rio de Janeiro, Brazil, June 1992." Port Louis, Mauritius: Government of Mauritius. Ministry of Environment and Quality of Life, 1991.

California Native Plant Society. "Rare Plant Status Report: *Lavatera assurgentiflora* (Kellogg)." May 1979.

————. "Rare Plant Status Report: *Lilium occidentale* (Purdy)." Compiled by Orrel Ballantyne and W. B. Critchfield. April 1978.

Center for Environmental Education. "Endangered Species Act Reauthorization Bulletin." January 18, 1982, no. 2.

Cochrane, Susan. "Why Rare Species?" *Outdoor California,* September/October 1985.

de Vos, Miriam P. "The Genus Romulea in South Africa." *Journal of South African Botany.* Supplementary volume no. 9. Published under the authority of the Trustees of the National Botanic Gardens of South Africa. Kirstenbosch, Newlands, Cape Province, October 31, 1972.

Duncan, W. H., J. F. Garst, and G. A. Neece. "*Trillium persistens* (Liliaceae), a New Pedicellate Flowered Species from Northeastern Georgia and Adjacent South Carolina." *Rhodora* 73, 1971: 244–48.

Farrar, Jon. "A Wildflower Year." *NEBRASKAland* magazine. 68(1), January/February 1990.

Figg, Dennis. "A Future for Endangered Species." *Missouri Conservationist,* April 1992: 12–17.

Fosberg, F. R. "The Deflowering of Hawaii." *National Parks & Conservation Magazine,* 49(10), October 1975: 4–14.

Hodge, Walter Henricks. "The Goliath of Seeds." *Natural History,* January 1949: 34–35.

Hynniewta, T. M., and P. K. Sarkar. "Orchids from a Conservationist's Eye." *American Orchid Society Bulletin,* December 1977: 1092–3.

Irwin, Howard S. "Miss Furbish's Lousewort Must Live." *Garden,* September/October 1977: 6–11.

Jain, S. K., and A. R. K. Sastry. "Threatened Plants of India, A State of the Art Report." Botanical Survey of India and Man and Biosphere Committee. National Committee on Environmental Planning and Coordination. Department of Science and Technology, New Delhi, India, 1980.

Jenkins, Dale W., and Edward S. Ayensu. "One-Tenth of our Plant Species May Not Survive." *Smithsonian*, January 1975: 92–96.

Jolly, Alison. "Island of the Dodo is Down to its Last Few Native Species." *Smithsonian*, June 1982: 94–100.

Kataki, Dr. S. K. "Indian Orchids—A Note on Conservation." *American Orchid Society Bulletin*, 45(10), October 1976: 912–14.

National Alliance for Plants. *Network*, Oakland, Cal.: April, August, and October 1980.

Sather, Nancy. "Western Prairie Fringed Orchid." Biological Report #36 of the Minnesota Natural Heritage Program Section of Wildlife, Minnesota Department of Natural Resources, 1991.

Schwartz, David M. "Where Have All the Wildflowers Gone?" *Country Journal*, April 1983: 50–56.

Schwegman, John E. "State of Illinois Recovery Plan for *Iliamna remota* (Greene)." Illinois Department of Conservation, January 1984.

Schwegman, John, and Bill Glass. "1986 Status Report on *Iliamna remota* in Illinois." Illinois Department of Conservation, 1986.

U.S. Department of the Interior, Fish and Wildlife Service. "Endangered and Threatened Wildlife and Plant Taxa for Listing as Endangered or Threatened Species; Notice of Review." *Federal Register*, 55(35), Wednesday, February 21, 1990.

———. "Family Lists of Candidate Endangered and Threatened Plant Species in the Continental United States." Excerpt from House Document No. 94-51, the Smithsonian Report. *Federal Register*, 40(127), Part V, July 1, 1975.

———. "Peter's Mountain Mallow (*Iliamna corei*) Recovery Plan." U.S. Fish and Wildlife Service, Newton Corner, Mass., 1990.

———. "Robbins' cinquefoil (*Potentilla robbinsiana*) Recovery Plan." Prepared by Joseph E. Doucette and Kenneth D. Kimball. U.S. Fish and Wildlife Service, Gorham, N.H., 1991.

———. "Small Whorled Pogonia (*Isotria medeoloides*) Recovery Plan." Prepared by Susanna L. von Oettingen. U.S. Fish and Wildlife Service, Newton Corner, Mass., 1992.

C O L O P H O N

THE illustrations for this book were drawn by Dugald Stermer, using graphite pencil and watercolors on Arches watercolor paper.

Designed by Samuel N. Antupit and Mr. Stermer, who were assisted by Miko McGinty, the book was set in Centaur and Arrighi types. Centaur, designed in 1912–14 by Bruce Rogers, reflects early roman letterforms used by the printers of fifteenth-century Venice, primarily the faces cut by Nicolas Jenson. While Centaur's antecedent is the stonecut letter, Arrighi is based upon the cursive written letter, which is attributed in its earliest typographic form to Francesco Griffo, a punch-cutter for the sixteenth-century Venetian printer Aldus Manutius. Arrighi, designed by Frederic Warde, was chosen by Mr. Rogers in 1929 to be the italic accompaniment to Centaur.

The text was composed and cast in lead on Monotype equipment; the introductory text pages at Mackenzie and Harris in San Francisco and the display and illustration commentaries at the foundry of Michael and Winifred Bixler in Skaneateles, New York. The copyright page was set in digitized Centaur and Arrighi on a Macintosh Quadra 650 by Ms. McGinty.

The book was printed by offset lithography at Dai Nippon in Japan from reproduction proofs of the type and from photographs of the original drawings by Color III Lab, San Francisco. The paper is Fukiage matte coated from Dai-Showa Paper Company, and the binding cloth is Asahi "T" Saifu.